D1047248

Successful ASIC Design the First Time Through

Successful ASIC Design the First Time Through

John P. Huber and
Mark W. Rosneck

Van Nostrand Reinhold
NEW YORK

Trademarks of the Mentor Graphics Corporation
QuickPath™ critical path analyzer
QuickSim II™ digital logic simulator
QuickFault II™ fault simulator
QuickGrade II™ fault grader
Design Architect™ design creation
NetEd™ Schematic drawing editor
SymEd™ Symbols drawing editor
System 1076™ VHDL design and simulation product
Genesil™ Silicon Compiler

Verilog is a registered trademark of **Cadence Design Systems**

Library of Congress Catalog Card Number 90-49999
ISBN 0-442-00312-9

Printed in the United States of America

Van Nostrand Reinhold
115 Fifth Avenue
New York, New York 10003

Chapman and Hall
2–6 Boundary Row
London, SE1 8HN, England

Thomas Nelson Australia
102 Dodds Street
South Melbourne 3205
Victoria, Australia

Nelson Canada
1120 Birchmount Road
Scarborough, Ontario MIK 5G4, Canada

16 15 14 13 12 11 10 9 8 7 6 5 4 3 2

Library of Congress Cataloging-in-Publication Data

Huber, John P.
Successful ASIC design the first time through / John P. Huber & Mark W. Rosneck.
p. cm.
Includes index.
ISBN 0-442-00312-9
1. Application specific integrated circuits—Design and construction. 2. VHDL (Computer hardware description language) I. Rosneck, Mark W. II. Title.
TK7874.H83 1991
621.39'5–dc20 90-49999
CIP

To Terri and Debbie, who stood by us through the late nights and long weekends to make this book possible.

Contents

Foreword xi
Preface xv

1. Introduction to ASICs and ASIC Design *1*

Introduction *1*
Doesn't Everybody Design with ASICs? *2*
How Many ASIC Vendors Are Needed to Build an ASIC? *3*
Types of ASICs *3*
The Benefits of Using ASICs *8*
When Should a Designer Consider Using an ASIC? *9*
ASIC Types: Advantages and Disadvantages *10*
When to Use a PGA, FPGA, or Standard ASIC *11*
ASIC Technologies and Trade-offs *11*
Market Statistics and Predictions about ASICs *14*
Summary and Conclusions *15*

2. The ASIC Design Process *17*

Introduction *17*
ASIC Design Process Flow *18*
Concept and Specifications: Where it All Begins *18*

The Foundry/Customer Interface 19
Design Implementation 24
Choosing an ASIC Vendor 24
ASIC Design Kits 27
ASIC Vendor Guidelines, Assistance, and Support 30
Design In-house, Rent, or Use the ASIC Vendor Design Center? 30
Postdesign Activities: Backannotation for Simulation 31
Summary and Conclusions 31

3. Design Creation for ASICs 33

Introduction 33
System on a Chip 33
Start the Design at the Architectural Level, From the Top 35
Behavioral-level Design 37
Functional Design 39
Structural Design 40
Detailed Design 40
Physical/Geometry 41
"Standard" Electronic Design Automation and Schematic Capture 41
Specific Design Considerations 44
Silicon Compilers and ASIC Design 52
ERC and DRC Considerations 54
ASIC and EDA Vendor Design Rule Checks 54
Knowing When the Design Is Ready for Verification 55
Summary and Conclusions 56

4. Design Methodologies for ASICs 57

Introduction 57
Bottom-up Design 58
Top-down Design 60
Conceptual-to-Functional Design 61
Design Synthesis 63
Logic Synthesis 66
Silicon Compilers 67

Should the Designer Lay It Out? 74
Summary and Conclusions 74

5. Logic Simulation 77

What Is Simulation? 77
The Imperative for ASIC Simulation 78
The Components of Logic Simulation 79
Initialization and Unknowns 91
Simulation States 92
Approximations and Limitations 93
Selecting the Computer Power 94
Tricks of the Trade 94
Acceleration Techniques 96
Developing a Successful Simulation Strategy 102
Top Ten Most Common Mistakes 104
Consider Test *Now* 106
Board-Level Simulation 107
Summary 107

6. Timing 111

The Importance of Timing 111
Basic Timing Considerations 112
Timing and Logic Simulation 114
Hazards 115
Specialized Timing Analyzers 118
Board- and System-Level Timing 124
Summary 125

7. Fault Simulation 129

What Is Fault Simulation? 129
Uses of Fault Simulation 129
Fault Simulation in the Design Process 130
Fault Simulation Basics 131
Inputs to the Fault Simulator 132
Outputs from the Fault Simulator 133

What Is a Good Fault Coverage? 133
Acceleration Techniques 134
Commonly Asked Questions 137
Some Straight Talk about Fault Simulation 138
Contractual or Government Requirements 140
Limitations of Fault Simulation 141
Summary 142

8. Test 145

A Story—Part 1 145
A Story—Part 2 146
A Story—Epilogue 146
The Moral of Our Story 146
Test and ASIC Design 147
DC Test Methodology 149
Design for Testability 153
Emerging Test Technologies 158
Power Analysis 162
Board Test 163
Summary 166

9. Summary and Conclusions 169

Introduction 169
What You Have (or Should Have) Learned—A Review 170
When to Use the ASIC Vendor's Design Tools and When to Buy Your Own 171
How to Choose an EDA Vendor 171
Long-Term Viability of ASIC Methodology for Board-System Design 173
Future of EDA Tools and Methods 173
Board and System Simulation 174
The Future of Mixed Analog-Digital ASICs 176

Glossary 179

Index 195

Foreword

The very name *application-specific integrated circuit,* or ASIC, connotes an ability to provide a dense package for a highly complex design targeted at a focused, often complex solution. The ability to create customized high-performance designs has come of age, facilitated by sophisticated tools that enable designers to cope with ever-increasing demands for added product functionality, features, and complexity. Most designers are trained in the traditional methods of approaching complex digital electronics with standard parts but have little, if any, exposure to custom or even semicustom integrated circuit design. Most see only a broad survey of IC technology. This book is targeted at the new ASIC designer who is getting ready to tackle that first ASIC design and is concerned about the unknowns that lie ahead. Economic and performance considerations as well as tool capability and process fabrication quality have evolved to the point where consideration of ASIC design is now commonplace in an ever-increasing number of electronic systems designs. Engineers are now given the challenge of coping not only with new technologies but with new design methodologies that are fundamentally necessary and advantageous to support new competitive high-tech products.

Laypeople and engineers alike have marveled at the advances made over the years in electronics' complexity, performance, density, and cost. The migration of systems to modules to boards to integrated circuits clearly underscores the radical transition that the physical incarnation of electronics has undergone. Engineers must also make a corresponding transition to cope with the physical implementation of creations whose parts and functions can no longer be easily subdivided or probed. The relatively high complexity that is packed into an integrated circuit today is often in striking contrast with the very limited ability of the designer to glean information from these tight little prizes of silicon.

The requirements on the designer and the design methodology he or she uses therefore become more stringent and more oriented toward creating a correct, accurate, and functional design from first fabrication. This need pushes the designer to be far more analytical and thoughtful in earlier stages of design to get his or her system-on-a-chip correct the first time through—perhaps not what experienced designers often do. Coping with design in the 1960s and 1970s, engineers often created brass boards or breadboards to evaluate key components, features, and functions. Once the physical implementation was created, the design would often be instrumented to determine whether or not it worked and how well it performed. The designer might even have the option, and sometimes the luxury, of optimizing a particular piece of the design. As design evolved in the 1980s and more and more automation was brought to bear on the design problem, for the first time engineers had the ability to determine whether or not a design would work before effort was expended to create the physical implementation. The designer still had to wait for the physical creation of the design and its integration into the larger system to cope with issues of optimization and detailed characterization. In the 1990s, the fabrication processes for ASICs have reached such levels of quality and complexity that developing large sections of electronics designs in any of a number of forms of integrated circuit technology customized to the application is now commonly given serious consideration. Likewise, design automation systems have evolved to the point where serious, detailed analysis of reasonably sized systems can and must be performed before committing to a density and package that precludes easy controllability, visibility, and even analysis of the design after fabrication. Yet these same methods and technologies yield unparalleled gains in cost, performance, quality, and reliability.

The processes applied to reap these fruits are well surveyed in the following pages. In Chapter 1, we are introduced to the dilemmas, trade-offs, and problems designers face in choosing among the myriad of technologies for inclusion in their products. The ASIC design process is reviewed in Chapter 2, and a detailed explanation of how to choose an ASIC vendor is given. Then we are introduced to state-of-the-art techniques for design capture and specifications in Chapter 3. Chapters 4, 5, and 6 provide a look at the approaches to synthesis, analysis, and timing that goes beyond the basics to offer design and usage techniques based on practical experience. The coverage of fault simulation and test in Chapters 7 and 8 introduces some of the more "sticky" design issues people face in attempts to cope with geometric growth in design complexities and verification. These surveys, colored with the wisdom of experience, offer designers new insights into what their own methodologies could yield.

Designers steeped in traditional approaches and methodologies toward electronics systems design have always needed to track changes in the technology, components, tools, and techniques of their craft. The inclusion of ASIC technology in their designs delivers such significant advantages that no designer today can honestly say that he or she is proficient and thorough at developing an electronics solution to whatever problem is being approached without an understanding and knowledge of the options afforded by the incorporation of ASIC technology. Training engineers and converting design methodologies to state-of-the-art approaches assisted by electronic design automation are becoming more and more necessary to competition in today's world markets. As with almost any human endeavor, experience begets new approaches, new techniques, and new questions to ask. Consideration of these new technologies, new tools, and new design methodologies molded by the experiences of many design iterations is surveyed herein. Not only is it useful to state-of-the-art systems engineers, design engineers, and engineering managers, it has now become a necessary part of the design process.

The information contained in this book will guide the designer through the realm of ASIC design, highlighting tried and proven techniques from which to choose for a new design. If you are designing your first ASIC or getting ready for another attempt and wish you could find some ready assistance with getting it done right with less pain, then this book is a must for you.

Dr. Geoffrey J. Bunza
Mentor Graphics **General Manager for Research and Development**

Preface

Effective digital ASIC design is accomplished by selecting the most appropriate technology and techniques to ensure the job is done on time on the first attempt. The single goal of this book is to make sure the right ASIC vendor and the appropriate tools are chosen so that a designer may develop the highest-quality ASIC in the shortest amount of time.

This book is provided for those design engineers who are about to engage in the development of their first digital ASIC. It also will be useful for those designers who are going to embark on their second visit into the realm of ASIC design and want to ensure success in less time and at a lower cost. There will be a detailed tour of the ASIC design process explaining how to find the right ASIC vendor, what kind of design tools are needed, the realm of design analysis tools, and how to measure the testability of an ASIC. This book assumes that the reader has a fundamental understanding of the integrated circuit development process, including wafer fabrication, mask development, and packaging techniques.

Mistakes in designing an ASIC are very costly, since an ASIC that does not meet its specification generally must be remanufactured. These "design turns" are extremely expensive. Many of these errors are due not to mistakes by the designer but to mismanagement of the overall design process. To avoid these costly miscues, this book will:

- Point out common pitfalls that inexperienced designers make and describe several shortcuts in ASIC design.
- Describe the ASIC design process in terms that a novice designer can understand.

- Make intelligent choices regarding the electronic design automation tools he or she might wish to employ.
- Describe the interrelationships between the designer, the EDA tool vendor, and the ASIC vendor.

The authors sincerely want you to be successful with your ASIC design experience. Therefore, they have attempted to reflect what is believed to be the consensus about the topics presented. This derives from the belief that if you stay in the mainstream for your first few ASIC designs, the chance of something catastrophic occurring is low. As you become more experienced, you may want to use some of the more advanced techniques described.

Many rules-of-thumb are presented. The intent is to give you a realistic target to shoot for and not to impose personal opinions on your design. Use these rules-of-thumb where they make sense. At the very least, they should give you some ideas about areas where you may want to seek the advice of a more experienced ASIC designer.

The authors have done their best to keep the reader in mind as they ordered their material and have continually asked themselves whether they were getting right to the core. This has led, in several cases, to decisions about the completeness of a discussion. Therefore, a philosophy was adopted that allowed a discussion of sufficient depth to permit you to make an informed decision about whether the technique being described is appropriate to your needs. Armed with this information you can discuss your specific requirements with your ASIC vendor and EDA supplier.

There are substantial rewards to effective ASIC design. Knowing how to design ASICs effectively can improve time-to-market, reduce costs, and eliminate many risks associated with ASIC development. We wish you good luck and much success.

The authors would like to extend their thanks to their colleagues at the Mentor Graphics Corporation. In particular we would like to thank Frank Binnendyk, Dave Hofer, Lauro Rizzatti, Kevin Harer, and Ken Salzberg for their expertise in reviewing the manuscript.

1

Introduction to ASICs and ASIC Design

INTRODUCTION

The application-specific integrated circuit (ASIC) has evolved from a simple array of a few hundred logic gates into a complete family of semicustom and full custom integrated circuit types using more than two hundred thousand logic gates. Semicustom ASICs offer preprocessed chips to which the designer only needs to add the final metallization connection steps. With full custom ASICs the designer builds the entire chip layer by layer, producing a truly one-of-a-kind integrated circuit. Because of the explosion in ASIC technology, ASICs are rapidly becoming fundamental building blocks for the design of boards and systems.

Some of the best reasons to use ASICs are the ability to reduce board space requirements, reduce development cost, increase reliability, maximize performance, and provide security for new designs. ASIC technology allows the integration of complex functions such as microprocessors and peripherals, coupled with memory, on the same chip. With the advances in logic synthesis techniques, ASIC design from concept to physical implementation has taken on new meaning for today's system designer. Designers who are not on the leading edge of ASIC technology may find their designs being beaten by competitive entries either in total performance or just by being quicker to market.

This chapter provides an explanation of the types of ASICs available, speculation about the ASIC market, and a discussion of ASIC technologies that are available to designers. Since this book is designed for the first-time ASIC designer, a lot of time will be spent on the basics of ASICs. After completing this chapter, the reader will know the different types of ASICs available, how to select the right type for a design application, the benefits of using an ASIC, and how to make the right technology choice, and will have had a brief look at market statistics on ASICs.

DOESN'T EVERYBODY DESIGN WITH ASICs?

No. Surprisingly enough, designs using ASICs are mostly located in large companies like IBM, Apple Computer, Hughes Aircraft, Boeing Corporation, Sun Computers, and Hewlett-Packard. Some industry experts believe that only 10 or 15% of all system designers have designed an ASIC. Since there are more than three hundred thousand system designers in the world, that leaves a significant number of designers who have not had the opportunity to experience the joys of ASIC design.

Large, successful companies have learned that using ASIC technology is a cost-effective solution to remaining competitive and meeting the challenges of today's complex designs. The benefits that these companies derive from using ASICs include reduced development costs, higher-performance designs, more complex designs, improved manufacturability, and higher reliability with new, complex designs.

Why aren't the rest of those four hundred thousand system designers designing ASICs? There are several reasons why ASIC design may be put off, ranging from not knowing how to begin, to (what is believed to be the most common) the costs involved. Initially, designing with ASICs is an expensive undertaking, especially since first-time ASIC design can be unforgiving. If mistakes are allowed to propagate their way into fabrication, the rework costs are twofold: increased development costs and increased time to market. There can be no white wires, no way to drill out pads or piggyback components. ASIC design is simply a matter of black and white with no room for gray areas. The designer must either do it right the first time or redesign and suffer the economic consequences, which can include missing a market window, extra development costs, and unreliable products. The most frustrating situation is that the designer cannot just quit, but has to redesign and pay for the privilege of redoing each step to, again, verify the fixes. This book takes novice designers through the steps that lead to designing an ASIC successfully, the first time. It leaves nothing to chance for first-time ASIC designers.

HOW MANY ASIC VENDORS ARE NEEDED TO BUILD AN ASIC?

Worldwide, the number of vendors supplying ASIC libraries is well over 150. Some of the top vendors are Fujitsu, NEC, LSI Logic, Toshiba, AT&T, Hitachi, Motorola, VLSI Technology, and Texas Instruments. The good news is that there is an ample supply of libraries with a variety of technologies from which to choose for developing an ASIC. The bad news is that there are so many vendors to choose from that it can take weeks or months to evaluate the vendors, the available technologies, and the offered libraries just to find what is right for a design.

These vendors are not limited to digital ASIC design. Full-function analog and mixed analog-digital technology ASIC support can also be found. Digital ASIC design support capability is by far the most popular, but providing highly complex analog functions integrated with digital-based structures is becoming a must. Most of the major ASIC houses offer digital, analog, and some level of mixed analog-digital capability with their libraries. Chapter 2 will provide the new designer with some guidance for the selection of an ASIC vendor.

TYPES OF ASICs

Each ASIC type can be implemented in a variety of ways, giving today's designers some help with making the right decisions in balancing costs against performance. Normal technologies such as CMOS, bipolar ECL, bipolar TTL, and gallium arsenide (GaAs) can be selected to provide an ASIC designer with the flexibility needed to make optimal trade-offs between density, speed, and power.

Though there can be a lot of debate, industry experts generally agree that there are *five* fundamental types of integrated circuit that are considered to be application-specific ICs:

Gate arrays
Standard cells
Compiled cells
Programmable logic
Full custom

The first four categories are considered to be ''semicustom'' ICs since only certain mask segments of the design process are controlled or can be customized by the designer. In the "full custom" category, the designer is responsible for the design of all mask layers, as many as thirteen separate process steps. A new capability in the semicustom cell-based area is the "megacell" or "supercell," which provides high-level functions for LSI- or VLSI-level integrated circuits, such as core microprocessors and peripheral support functions.

Even experienced system designers are learning new things about the tools and technologies of the rapidly evolving area of ASIC design. With each passing year, new capabilities appear that offer additional ways to pack more functions onto a silicon substrate or to squeeze a few more megahertz from an old design. An old adage that can sum up the ASIC opportunity is, "Today's boards are tomorrow's ASICs."

The new ASIC designer must know and clearly understand the fundamental differences between the types of ASICs available. The designer must then be able to choose the ASIC type that will be best for the design.

Gate Arrays

Gate arrays (GAs) consist of preprocessed wafers of logic elements (gates) that require only between one and three mask steps of metal interconnect to complete the fabrication process. The GA structure usually consists of columns of transistor arrays that will be configured to form basic logic functions chosen from a cell library, and are surrounded by I/O pads. The designer simply chooses which logic functions will be needed and how the basic logic functions will be connected on the chip. The ASIC vendor will process the chip by using layers of metal to connect the logic gates with each other and the output pads to complete the array. Part of the gate array includes power and ground buses that can be connected to I/O pads to provide power to the chip.

GAs can provide designers with more than two hundred thousand logic gates on a single chip. However, because of difficulties in creating signal routing channels, only between 70 and 90% of the available logic gates are usable. With this large array of available logic gates, GAs force the designer to pick a preconfigured I/O structure, which can limit design flexibility and force the designer to make trade-offs that would not be necessary with another ASIC type.

Gate arrays can be packaged as dual-in-line packages (DIP), leadless chip carriers (LCC), and pin-grid arrays (PGA). Packages can have from sixteen to over three hundred pins allowing designers to pack the silicon with a considerable number of functions, although designers have to pay for the extra gates and silicon that they are not going to use or cannot use.

Newer techniques are being used to provide more usability with large gate arrays. The most popular is the "sea-of-gates" architecture (see

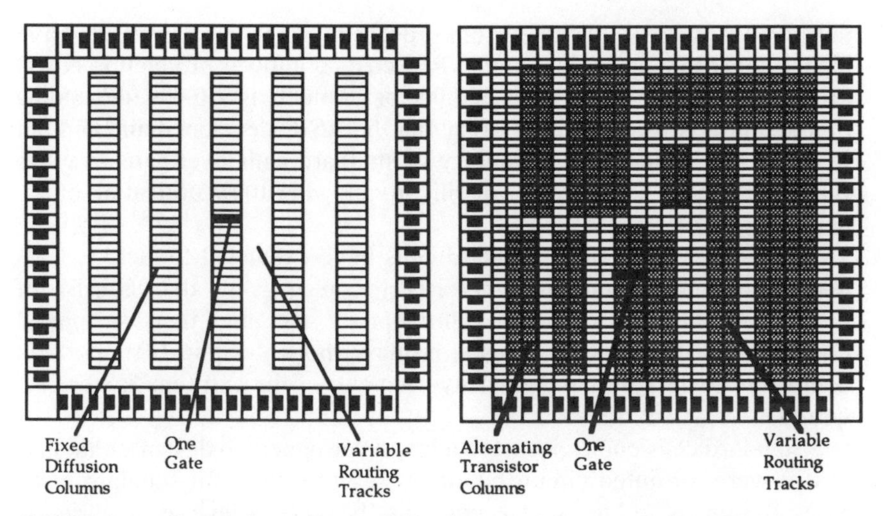

FIGURE 1-1. Comparison of a traditional channeled array (*left*) and the LSI Logic channel-free array (*right*). *Courtesy of LSI Logic Corporation 1990.*

Figure 1-1), which allows connectivity routing over unused logic gates and eliminates specific routing channels. By eliminating dedicated routing channel areas, gate arrays can be made to handle denser architectures. The sea-of-gates architecture provides much larger densities, to 250,000 gates, but only about 40% of the gates are usable.

Gate arrays were first used to handle the "glue" logic in digital board designs, allowing designers to reduce the need for discrete digital components. Gate arrays are still very popular for board design but are slowly being replaced by standard cells.

Gate arrays are used when a quick turnaround is needed for prototypes or when the production volume is less than ten thousand units. Gate arrays are simple, offer quick turnaround (but not as fast as FPLD or PLD), and provide a means of implementing large designs. The major drawback with using gate arrays is the lack of the flexibility to add complex functions, such as standard microprocessors.

Standard Cells

Standard cells (SCs) are made up of predesigned circuit functions at the LSI/VLSI level of complexity that can be joined by interconnecting cells. SCs are cheaper to use than GAs if the unit volume is greater than ten thousand units but less than about one hundred thousand. Standard cells offer more flexibility but require more custom mask processing steps for fabrication. Because of the custom processing steps, standard cells have higher nonrecurring engineering (NRE) costs but lower production costs. NRE, which can be as little as $20,000 or as much as $50,000 (depending on the chip), is the basic cost of having the ASIC developed and laid out by the ASIC vendor. An advantage of standard cells over gate arrays is that there are no wasted gates or silicon, allowing 100% utilization of the chip area.

In the last few years, standard cells have expanded to include LSI-level components termed *supercells* (or megacells) that consist of complex functions like popular microprocessors and their peripheral components. Cell-based designs require more technical support to complete the design than gate arrays because of the extra mask steps required.

Standard cells come closest to allowing designers to design a chip just as if it were a printed circuit board. A key to successful standard cell–based designs is to choose a vendor that has a high level of experience with standard cell designs. Two important indicators of a vendor's experience are large libraries of precharacterized transistor layouts and revenue history.

Compiled Cells

A compiled cell is a function that has been derived using a set of parameters with a silicon compiler to provide either a custom or a standard function cell. Compiled cells are similar to macrocells from a semicustom standard cell IC library, but usually require more die area. Application-specific compilers are available to build such complex functions as RAM, ROM, PLA, and multipliers. Normally, compiled cells are derived directly from a schematic and compiled into physical geometries for layout. Chapter 3 will provide more detail about silicon compilers and the resulting cells.

Full Custom

A full custom IC is an ASIC that is designed without using any precompiled or preprocessed silicon. The custom IC is handcrafted at the transistor level by the designer to optimize each cell for area and performance. Custom ICs can have a higher transistor density per chip than other types of ASIC because each cell is hand placed. This makes development of a custom IC a more time-consuming project since all mask steps must be completed. The number of mask process steps is usually thirteen, including metallization with one, two, or three levels of metal interconnect. The major benefits of a full custom IC design over the other ASIC types are that the density is higher, there is no wasted space, and the production costs can be much lower than for other types of ASICs.

The density is the highest because of the manual transistor and cell packing techniques used in the development process. All transistor-based cells are hand built (not pulled from a predesigned cell library) and manually placed on the chip to produce the maximum density possible. Because of this manual packing, NRE costs are much higher for full custom IC designs, but unit volume costs can be the lowest, especially if large quantities are needed (usually more than one hundred thousand). If time to market is an issue, keep in mind that a full custom IC will require considerably more time to complete than a standard semicustom ASIC.

Programmable Logic Cell Gate Arrays

Programmable gate arrays (PGA) and field-programmable gate arrays (FPGA) are actual gate array parts without any connections. However, logical function blocks can be programmed on the fly by using configurable logic blocks (CLBs) to connect the basic functions in a desired manner.

A PGA is simply a predesigned chip with groups of cells that are configured as AND-OR arrays with their outputs feeding a flip-flop. A good rule of thumb to use for measuring the size of the PGA is to look at the number of flip-flops it contains. PGAs can have as many as 128 flip-flops, with the amount varying depending on the vendor used. There are many different names for programmable logic arrays (PLAs), including programmable logic devices (PLDs) and programmable array logic (PAL).

FPGAs differ from programmable logic arrays in that where the PLA has a sum-of-products arrangement of logic components, the FPGA has a core of logic cells surrounded by I/O cells. The major drawback is that the number of gates available is less than twenty thousand. The major benefits are that no mask steps are required to configure the FPGA, and it can be used repeatedly for different applications simply by reprogramming the CLBs. Logic functions can be selected by programming a RAM-based function generator module contained in a configurable logic block, sometimes referred to as a CLB.

PLA/PLD ASIC designs consist of an array of sum-of-products gates that are programmed through equations input into a standard PLA/PLD program. In recent years, flip-flops and latches have been included in the collection of unconnected gates, expanding the use of these ASIC devices. These devices offer the quickest turnaround for prototyping. However, they do not provide the speed or the density that an equivalent gate array would, and therefore they cannot provide a complete model to verify product specifications. A designer may elect to use a PLD architecture to get quick prototypes in order to begin verifying a new design. After the verification is completed, the new ASIC can be transferred into a cell-based or array-based architecture for manufacturing production.

THE BENEFITS OF USING ASICs

The benefits derived from using ASICs in board designs fall into five major categories. Each of the listed benefits by itself may not sell engineering management on using an ASIC in the next new design project. However, all of these benefits taken together make the use of ASICs very economical and practical. If we add to this the fact that ASICs offer a competitive advantage in today's highly technical fields and competitive electronics industry, we can come to the conclusion that ASICs are a must for today's and tomorrow's system designs.

- **Achieve additional complexity in silicon.** ASICs allow more complexity to be included on a single silicon substrate. This means that complex functions, requiring several hundred thousand transistors, can be fit onto a single silicon die using standard cells of core microprocessors, RAM, ROM, and I/O. This is similar to miniaturizing a board design. It has long been stated that toaay's boards are tomorrow's ASICs.

  This complexity does not come without a cost. The smaller the chip size desired, the longer the design will take and the more it is likely to cost.
- **Retrieving or providing additional board space.** Using ASICs on a board will allow space that would normally have been used for "glue" logic to be freed. With today's technological advances in ASIC design and the functions offered, functions that required two or more components can now be combined into one. This reduces the total parts count as well as space and power requirements.
- **More capability for less cost.** ASICs allow a designer to put more functions into a smaller space, reducing the parts count and allowing boards to be reduced in size or allowing functions that previously fit on two boards to be placed on one.
- **Reduced development time.** ASICs have a much shorter development time than board designs of the same complexity. While the initial design conception, schematic creation, and simulation steps remain the same, the fabrication and test steps are reduced since all the components are built on a single piece of silicon.
- **Additional reliability**. Board design reliability goes up as fewer parts are placed on a circuit board. This reliability continues to improve

as a lower parts count contributes to a reduction in connections and traces on the circuit board.

WHEN SHOULD A DESIGNER CONSIDER USING AN ASIC?

ASICs began as quick and easy substitutes for standard TTL logic, known as "glue" logic for printed circuit boards, as far back as 1976. The first to adopt ASIC technology were the military and high-speed computers and telecommunications industries, who used ASICs to meet their stringent density and performance needs.

When designers need to reduce the size of existing printed circuit boards or replace large sections of TTL logic, their best choice is to explore ASIC solutions. The more appropriate questions are, Where should I use an ASIC? Which type should I use? Which technology should I choose?

To answer these questions, designers must understand the ASIC design process and the tools that are used to effectively perform the tasks. Most ASIC vendors feel that there are six basic development steps for an ASIC design. These are:

Schematic capture

Simulation

Package assignment

Physical layout

Back annotation and resimulation

Test vector generation

One could argue that there are also three more steps: prototype manufacturing, prototype test and verification, and ASIC production. In this book we will primarily concentrate on the basic six steps but discuss the additional three as required.

Each ASIC vendor provides guidelines about which steps can be performed at the designer's local facility and which steps must be performed at the ASIC vendor's design center. The ASIC vendor should provide a design kit that will outline each of the steps required and who is responsible for each. Normally, an ASIC vendor will supply ways to help new designers specify design information about the ASIC's size, package type, and bonding. The ASIC vendor will expect the initial ASIC specifications to have been defined at the designer's facility before a design is submitted to the vendor. Chapter 2 will explore the relationship with the

ASIC vendor and detailed ASIC requirements and specifications in more detail.

Throughout the remaining chapters of the book, we shall explore in detail what is involved in each of the design steps for an ASIC.

ASIC TYPES: ADVANTAGES AND DISADVANTAGES

This section will examine the differences between full custom ICs, standard cells, and gate arrays. After the NRE charges, the package pin count is most often the primary contributor to the cost of an ASIC, even over the number of gates. The costs of an ASIC will vary based upon the quantity ordered, with price breaks showing up at 100–1,000, 1,000–10,000, and 10,000–100,000 and up. Of course, the larger the quantity, the lower the cost.

Gate arrays are the easiest to implement, have the faster turnaround time for prototypes, are easiest to change, and can be the cheapest to implement in low volumes. Normally needing only one to three mask steps to add connective metal layers for complete fabrication, the gate array requires very little effort to implement compared to other semicustom types. In addition, with the sea-of-gates approach to gate array technology, the density can be as much as 250,000 gates per chip. The standard gate array can offer densities in the 20,000–50,000-gate area with up to 95% utilization. The sea-of-gates array will provide gate utilization of only up to 40%. The advantage of the sea of gates is that 40% of 250,000 gates is 100,000 gates, which is still twice the gate density of a standard 50,000-gate design. This high density paves the way for more on-chip RAM, which allows faster access times to aid system designers. The gate array allows only simple logic gates, and there is no flexibility to modify the width or height geometries of devices on the substrate. The gate array is still primarily used to replace glue logic in printed circuit board designs.

Standard cells are the next easiest to implement but may require as many mask steps as a full custom chip. The biggest advantage of standard cells over gate arrays is that they can be more densely populated. Standard cells have more flexibility in device geometries since one dimension, the height, is variable. Gate array geometries are fixed in height and width and therefore are limited in the number of logic gates they can support. Standard cells require up to thirteen mask steps for complete fabrication. They can require anywhere from two to four times the fabrication time needed to produce a gate array.

Full custom is the most complex to implement and the most costly. Full custom requires all thirteen mask steps to complete the fabrication. The designer must customize each mask step to provide maximum densities, maximum functionality, and maximum performance for each device. A full custom ASIC will require the most time to develop and will incur the highest nonrecurring engineering (NRE) costs. The full custom approach allows the designer to incorporate the highest level of function complexity on a single chip. The designer must determine whether to make one large chip or to segment the design into multiple chips to distribute the complexity.

The final choice about which ASIC methodology to use revolves around the trade-offs between two issues: production volume and nonrecurring engineering costs. Table 1-1 shows a comparison of NRE with production costs per unit for the three main ASIC types.

TABLE 1-1. Trade-offs for Costs among ASIC Types

NRE	*Production Costs/Unit*
Low: gate arrays	Low: full custom
Medium: standard cells	Medium: standard cells
High: full custom	High: gate arrays

WHEN TO USE A PGA, FPGA, OR STANDARD ASIC

The choice of whether to use a PLD rather than an ASIC or standard gate array can be difficult. Often a PLD is chosen for the development of initial prototypes, and then a standard ASIC is used for final production. This can work only if the ASIC design is fully simulated in its complete and final implementation in its target board environment; the ASIC is verified as plugged into the board for simulation. Either way, the designer must study the specifications of the PAL/PLD to ensure that the desired results can be obtained without using the full ASIC in the design.

ASIC TECHNOLOGIES AND TRADE-OFFS

There are several popular technologies available for ASIC design that can fit every need in a board or system environment. The choice is up to the designer, though the choice need not be made until the design is ready for fabrication. As will be discussed in Chapters 3 and 4, many ASIC design trade-offs need not be made until later in the design cycle. The designer may even want to make different technology versions of the same design to satisfy both commercial and military needs. Technology selections fall into five categories: CMOS, TTL, ECL, GaAs, and BiCMOS. It is expected that the reader is familiar with the fundamentals of TTL, ECL, and CMOS logic technology, so only the newer technologies will be described in detail.

CMOS is by far the most popular technology in use today. Its main advantages have been the density that can be achieved and its low power dissipation. The major drawback of CMOS has been the operating speed, but lately CMOS has been gaining in performance against bipolar TTL devices. Bipolar technology ASICs have been perceived as one-dimensional, supplying only higher clock speeds and larger drive capabilities than CMOS. CMOS has been gaining in the operating speed area but not the output drive. There is a solution available, however, that covers designers' needs. The recent trend has been to use CMOS for all of a cell's internal logic functions, but to use bipolar technology for I/O buffer/driver applications in the same cell.

This arrangement, called BiCMOS, has enabled designers to get the desired density and logic functionality required along with the drive characteristics needed to interface to external devices. BiCMOS technology allows ASIC designers to have both the density and low power of an all-CMOS design and the speed and drive power of bipolar.

The melding of the technologies is shown in Figure 1-2 using a two-input NAND cell example. Notice how the MOS transistors are used for handling the logical NAND function while the bipolar transistors are used for output drivers.

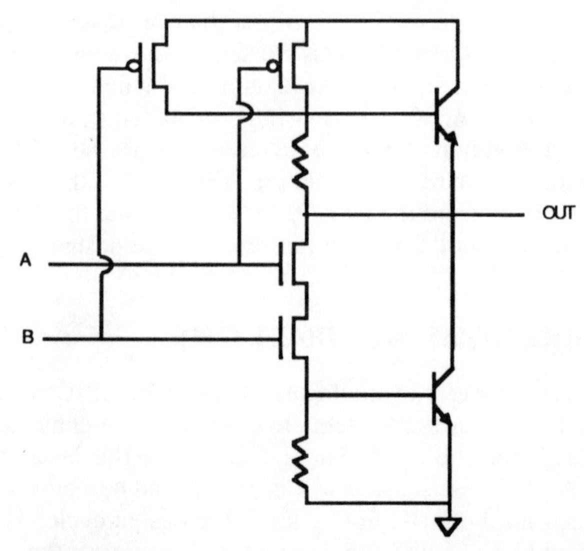

FIGURE 1-2. BiCMOS two-input NAND gate cell example.

BiCMOS provides an ideal combination for most designs from low speed (less than 10 MHz) to clock speeds of up to 50 MHz. BiCMOS's other benefits compared to bipolar are:

- Faster switching speeds
- Power reduced by up to 25%
- Current (active) cut by 50%

With the drive to shrink device sizes, there is very little use for TTL-only ASIC devices. However, there is still a need to achieve higher-speed devices. CMOS can provide satisfactory clocking for digital circuits up to 50 MHz. Beyond the 50-MHz barrier, the designer must turn to either bipolar ECL or gallium arsenide technologies to achieve the performance desired.

ECL is used where faster speeds are needed with little regard for power dissipation. There are some cell-based libraries for ECL arrays, making them a more attractive alternative for high-speed products. The ability to use process technology between 0.8 and 2 μm and gate counts to 100,000 gates continues to provide ECL with a large following. The density is still well behind that available with CMOS technology, and it does not seem likely that ECL will ever catch up. Of course, when the circuit speeds up, power dissipation concerns arise, especially when using ECL. If a designer uses ECL, the device may require an elaborate scheme to cool it. Several manufacturers offer complex heat sink additions for larger chips in the 30,000-and-over gate range, using over 400 pins and dissipating up to 30 W of power. ECL can offer clock speeds to greater than 1 GHz, with 100–250-ps gate delays. Typically, ECL exhibits 50–60 ps of delay for every millimeter of interconnect.

If speed is the primary requirement, the designer has only two choices, ECL or gallium arsenide (GaAs) technology. The speed requirement also brings along a power dissipation constraint on the package.

Gallium Arsenide (GaAs), the latest high-speed technology that is rapidly coming of age, uses enhancement/depletion mode gallium arsenide instead of silicon as the chip substrate. GaAs is slowly closing the gap on ECL to rival the long-standing high-speed technology leader. The 1990s will see GaAs provide the right density coupled with the high speed (gigahertz) needed to support designers' needs for high-performance computers and telecommunications capabilities.

GaAs offers several advantages over ECL for high-speed ASICs. These advantages are:

- One-fourth the power dissipation
- A simpler device structure
- Voltage swing of 500 mV

- Improved noise margins
- Fewer fabrication steps

GaAs has much more stringent design constraints because of its higher clock speeds. The most obvious concern is with clock duty cycles where signal rise times are slower than signal fall times. This problem causes pulse distortion, sometimes known as pulse swallowing, which can be a concern for synchronous clocking designs.

TABLE 1-2. ASIC Technologies and Their Key Characteristics

Technology	*Speed*	*Density*	*Power*
CMOS	50 MHz	Highest	Lowest
TTL	100 MHz	Medium	Low
BiCMOS	50 MHz	High	Low
ECL	>1 GHz	Low	Highest
GaAs	>3 GHz	Lowest	Medium

MARKET STATISTICS AND PREDICTIONS ABOUT ASICs

ASICs are going to continue to be a popular item around the world. However, industry analysts are making only conservative predictions about the market growth for 1991 and beyond. The estimated overall market compound annual growth rate is expected to be only about 10% in 1991.The three main market segments tracked are semicustom, custom, and programmable logic devices (PLD). PLD is now being tracked as a separate segment from semicustom due to its explosive growth over the past year.

Gate arrays are expected to continue to be the most popular means of making an ASIC. The most explosive growth area was in the field programmable gate array (FPGA) area where the market size doubled in 1990 and is expected to grow about 45% in 1991. Linear gate arrays are likely to see a modest growth over the next 3 years.

The two reasons for the tremendous growth of FPGAs are the reduced time of market and the low NRE costs for the FPGA over standard gate arrays. Major deals among ASIC vendors allowed FPGAs to be converted to gate arrays and provided a means of lowering customer development risks.

Technology

In the area of technology, analysts agree that Bipolar use is going to hold its current run rate or see a slight decrease over the next 3 years. BiCMOS use, in contrast, is expected to grow over 80% through 1991 but this rate will slow to about 55% by 1993. BiCMOS should continue to produce devices with higher pin counts that will have operating speeds edging toward the 100Mhz area. As BiCMOS continues to grow, pure bipolar devices use will continue to shrink.

Gallium Arsenide (GaAs) is most certainly an ASIC technology to watch closely. GaAs ASIC use was up 32% in 1990 and is expected to give both ECL and CMOS a challenge during the next several years in speed and density. GaAs is expected to offer capabilities in the over–100,000 gates range and even as high as 350,000 gates in 1991 with up to 70% gate utilization. With these kinds of gate densities, GaAs will clearly be making a run on the bipolar ECL market and should even give CMOS a challenge. Designers should take a close look at the GaAs offerings and do some comparative pricing. GaAs is expected to compete head-on with silicon base ASICs in pricing. Some ASIC vendors have announced new offerings that are a price/performance match for BiCMOS and priced lower that comparative ECL arrays.

Analysts believe that ASICs will make up between 14% to 16% of the worldwide IC market. What all this market information means to new ASIC designers is that there will continue to be plenty of suppliers, an expansive array of products, and more attractive pricing to help offset development costs. Additionally, it means that system design using EDA tools will get easier as they keep pace with ASIC design requirements.

All of these advances will encourage EDA vendors to provide the tools that designers will need to design, verify and test new ASIC-based products.

SUMMARY AND CONCLUSIONS

ASICs have penetrated every segment of the electronics industry. The gate array is being replaced by the standard cell as the basic building block for new designs. Larger standard cells called megacells or supercells are bringing IC design into the realm of designing a board. This is being accomplished by using core microprocessors and peripherals component functions on a common substrate.

Programmable gate arrays are seeing increased use as the number of usable gates grows sufficiently large to handle a designer's needs, especially for prototyping. New trends point to an increasing use of mixed analog and digital components on a die.

REFERENCES

ASICs Defined. *ASIC Technology and News,* May 1989.

Dugan, Thomas D. GaAs Gate Arrays. Vitesse Semiconductor Corp., *High Performance Systems,* February 1989.

Fawcett, Bradly K. Taking Advantage of Reconfigurable Logic. Xilinx Inc., *Programmable Logic Guide,* 1989.

Gabay, J. A Burst of Activity on Sea-of-Gates Arrays. *Semicustom Design Guide,* 1989.

Gabay, J. Programmable Loci Architecture: From PLDs to FPGAs.

Gheissari, Ali, Programmable Macro Logic: New Solutions for Designers. Signetics Co., *Programmable Logic Guide,* 1989.

Kopec, Stan, High Level Approach Speeds Design of Complex PLDs. Altera Corp, *Programmable Logic Guide,* 1989.

McCarty, D. "Compare Architectures, Not Density Claims, When Evaluat Field Programmable Logic Devices." *ActToday* 2, no. 1 (1990).

Meyer, E. ASIC Users Shift Analog On-chip. *Computer Design,* March 1990.

Meyer, E. Bicmos, ECL and GaAs Chips Fight for Sockets in 1990. *Computer Design,* Jan 1990.

Runner, J. Scott, The Next Step Upward in ASIC Functionality: Supermacros. Fujitsu Microelectronics Inc., *Semicustom Design Guide,* 1989.

Semicustom Design Guide, VLSI Systems Design, 1988.

Sharp, D. and G. Barbehenn, Squeezing State Machines Into PLDs. *VLSI Systems Design,* October 1989.

2

The ASIC Design Process

INTRODUCTION

Designing an ASIC is becoming similar to designing a printed circuit board. Specifications are developed; then components are selected, connected, and tested. Today, an ASIC designer can begin assembling the chip by choosing the functions needed from an extensive functions library. ASIC vendors offer specific design methodologies based on the manufacturing processes that they use. Compilers are available for regular structures, such as memories, multipliers, and datapaths. Currently, each ASIC vendor offers a wide range of processes for the designer to choose from. Which choices the designer makes will ultimately determine how the ASIC will perform, the size of the package, the implementation technology used, and how much it will cost, both to develop prototypes and to manufacture.

The ASIC design process is broken into a sequence of stages. Each stage must be successfully completed before the next begins to avoid unnecessary redesign and NRE charges. Each stage may differ in length depending upon the complexity, the technology, and the designer's experience. Figure 2-1 shows a simple example of a generic ASIC design process. The basic, and therefore simplest, ASIC component to develop is the gate array. The main reason that the gate array is the easiest to design is that most of it is already designed; only the final metallization steps (one to three layers) need to be completed. It is interesting to note

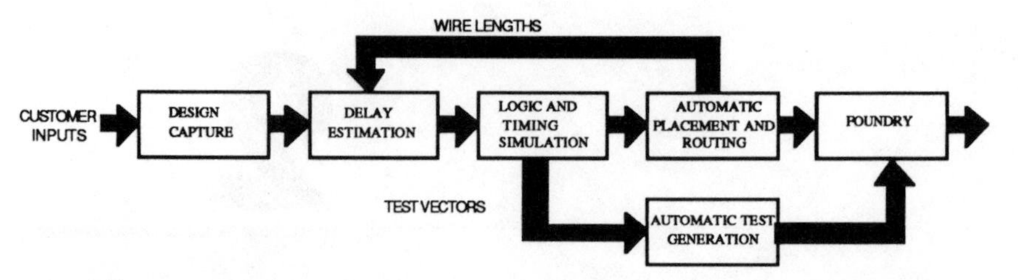

FIGURE 2-1. A generic ASIC design flow depicting tools at the ASIC vendor.

that the process will vary depending upon the vendor used, whether EDA tools are used or not, and, again, the complexity and technology of the design and the experience of the designer.

ASIC DESIGN PROCESS FLOW

The process flow will be described by detailing each discrete step. The approach will be top-down, beginning at the conceptual stage (labeled customer inputs) and progressing through the detailed chip specification, the actual design, and the preparation stage for analysis. The "back-end" process steps, simulation and test, will be thoroughly described in Chapters 5, 6, 7, and 8. This book will examine the layout stage only briefly since layout is normally performed by the ASIC vendor; the designer has little to contribute during this stage. However, we will touch on different aspects of the layout process that the designer can influence to reduce costs, reduce risks, and eliminate the need for a second layout.

CONCEPT AND SPECIFICATIONS: WHERE IT ALL BEGINS

The first step in a new ASIC design is to define the functional specifications, which lead to identifying the characteristics that the ASIC must have to fulfill its role. This first step is one of the most important since the remaining process steps will derive their inputs and guidance from these specifications. Whether the new ASIC is being conceptualized on a napkin or an EDA workstation, the concepts and specifications step must be carefully worked through. Shortcuts in this step could have a substantial impact on developments throughout the later stages of the new ASIC's design. Industry experts have found that the decisions made in the first 5% of the design cycle establish 85% of the total development costs. The point here is to put more effort into the definition stage, early in the design cycle, to avoid unpleasant and very costly surprises later.

The conceptual stage is approached with some idea of the ASIC's primary function. This may simply be to replace several thousand logic gates on a printed circuit board, or it may be to provide a new microcontroller function for the I/O stage of a new design. Usually, the type of ASIC to be used will be determined from the specification stage.

ASIC vendors provide new design information checklists to help in defining new designs. Table 2-1 shows some of the key requirements that

TABLE 2-1. Design Specification Information for the ASIC Vendor

Specification	*Defined*
AC characteristics	Maximum frequency
Power	V_{CC}, V_{DD}
Packaging	DIP, PGA, LCC, QFP
Gate count	Total number of logic gates used
Buffer count	Number of buffer gates
I/O	Number of I/O pads
Pin count	Total number of pins

should be specified for a new ASIC design *before* the design is implemented. This detailed information will allow the vendor to help the designer identify the processes and technology options that are available for the new design. As can be seen from Table 2-1, detailed constraints on the design must be specified at the beginning of the development process.

The conceptual stage must be performed well in advance, before going to an ASIC foundry. The best approach for a first-time designer is to perform conceptualization, system requirements, and logic design using technology-independent libraries, then do an ASIC vendor search. The reason for this, though it may be obvious, is to ensure that the designer can determine whether the ASIC vendor has the experience, the capacity, the technology, and the support available to undertake the new ASIC design successfully. If the ASIC vendor has no track record in the specific technology required, problems may develop later in the design cycle, causing additional time to be needed for design completion.

THE FOUNDRY/CUSTOMER INTERFACE

Choosing a vendor for simple gate arrays is becoming more difficult even though there are fewer vendors than there were five years ago. The difficulty is that many vendors offer the same library capabilities with the same device geometries, gate/propagation delays, densities, architectures, and process choices. The designer's choices for a vendor providing standard cell libraries are similar. Most vendors even offer higher-level functions, such as microprocessors and peripherals with on-chip memory arrays and mixed analog-digital cells.

What seems to hold true for all ASIC vendors is that the most critical decisions in choosing a vendor are made well before the designer begins to lay out the circuit. Factors that can influence the design process and which vendor to use are:

Production capability/capacity/track record

Turnaround time

NRE costs

AC characteristics

DC characteristics

Packaging

Functions/functionality

Support—local design center/applications engineering

Vendor and customer preconceptions

Quality assurance

The designer must have a clear understanding of each of the factors that will influence design decisions. This understanding is critical to a successful ASIC design the first time through. This section provides a definition of each of these factors.

Production capability/capacity/track record is the vendor's ability to deliver what the customer needs, when it is needed, in the quantities required—all proven through experience. It is a good idea to determine whether the vendor manufactures the ASIC directly or uses a third party. If a third party is used as the manufacturing agent, then a thorough check of the third party's manufacturing ability and reputation is required to avoid any unpleasant misunderstandings should anything go wrong. The designer should determine what the support policy is and whether the ASIC vendor will take responsibility for any problems. Ask for references if in doubt as to the vendor's ability to deliver.

Turnaround time is the measure of how long it takes the vendor to deliver *first* prototypes *and* quantity production parts. Turnaround times are measured from the time when the customer first delivers to the vendor a verified netlist and simulation test vector files to the time when the vendor delivers the set of fully tested prototypes. Turnaround times can be anywhere from six to twelve weeks for standard cell–based designs and one to three weeks for array-based designs.

NRE and other development costs can cause the design to exceed the initial budget, affecting the customer's planned profit targets, and must, therefore, be carefully tracked. Designers must investigate the production price variance for different processes and technologies, computer time costs, computer requirements, and NRE that may be hidden or even negotiable. NRE can be as much as $150,000, so designers are encouraged to find out which services are considered "extras." For example, design rule checking may be an extra, or there may be a time limit on use of the mainframe computer, after which you are charged for additional CPU time.

AC characteristics are most often related to clock speed and propagation delays. Designers must consider both specifications to ensure proper operation with chosen technologies. Often an ASIC design fails because only the raw silicon speed of the device is taken into account, not additional delays related to channel lengths or interconnects. Failure to include propagation delay tolerances can lead to higher manufacturing costs and costly delays in getting the ASIC into production. Generally, propagation delay is measured based on the standard input-to-output delay of a two-input, one-output NAND gate.

If the design is to be mixed analog-digital, special attention must be paid to parasitics. This is especially a problem if the process parameters for the analog and the digital differ widely, such as 2-μm analog with 0.8-μm digital. Analog amplifiers, sensing circuits, and filters are very sensitive to parasitics, which can cause difficult design problems if not handled through proper initial planning.

By taking advantage of cell structures and paying closer attention to critical paths, a design can be made more tolerant of technology process restrictions (such as 2.0 μm instead of 1.2 μm) to keep costs under control. This attention to detail will provide an ASIC that is easier to design, test, and manufacture, and also provide a more reliable and more inexpensive part. The designer can always advance up the technology chain as the market demands increase.

DC characteristics pertain to dc drive current requirements for capacitive loading, distributed clock signals, and memory drive. DC requirements should be specified for every I/O pin on the chip. These specifications should include detailed requirements for leakage current and source and sink currents.

DC characteristics affect on-chip power requirements and must be calculated and checked carefully before the design is released to the vendor to keep package costs low and manufacturability high. This means that the designer should define maximum and minimum power-supply current for each supply in the system, including maximum power dissipation. Power-supply variations should be clearly specified and should include temperature ranges and variations.

Packaging considerations have to do with whether the part should be surface-mounted as a dual-in-line package (DIP), leadless chip carrier (LCC), quad flat array (QFA), or pin grid array (PGA). This will affect whether the package must be manufactured using a plastic or ceramic housing. The packaging consideration is often left to last, when it is too late to make fundamental changes. Decisions about packaging pertaining to surface mount, sockets, or through-hole soldering for both prototype and production will affect overall cost, manufacturability, and field servicing.

Functions/functionality pertains to the number of functions that the ASIC must provide. The functionality will determine the silicon size, the ASIC vendor, and the EDA environment needed to complete the design. The requirements for functionality will also directly determine the die size, testability, manufacturability, and cost.

Functionality for both current and future design requirements should be considered. For example, the designer may partition a design into multiple ASICs rather than using a single chip. The multiple-chip scenario may offer the benefit of allowing a set of functions in the design to be reused in other designs with minimal change or tooling costs.

Support–local design center/applications engineering should be checked out to determine how support will be handled. Knowing how and from where applications engineering support will be available and to what extent is a must for a first-time ASIC designer.

Vendor and customer preconceptions cover the areas of how the customer-vendor interface is handled (see Figures 2-2 and 2-3). Who is doing the design, and what is the experience level? Developing guidelines for internal designers to follow when designing an ASIC is a worthwhile exercise that will produce early results in cost savings and increased productivity. To further ensure against design iterations in silicon, the new ASIC designer should be able to provide the ASIC vendor with information about the target environment. The information needed here concerns external devices that may drive or be driven by the ASIC and is to be used for selecting the right macrocells from which to develop the ASIC.

Quality assurance pertains to the actual manufacturing and verification steps that the vendor performs to ensure that the new ASICs function as specified. The first-time ASIC designer should take extra time to understand the process steps that will determine how well the new design will function and what actions can be taken if it does not work as expected.

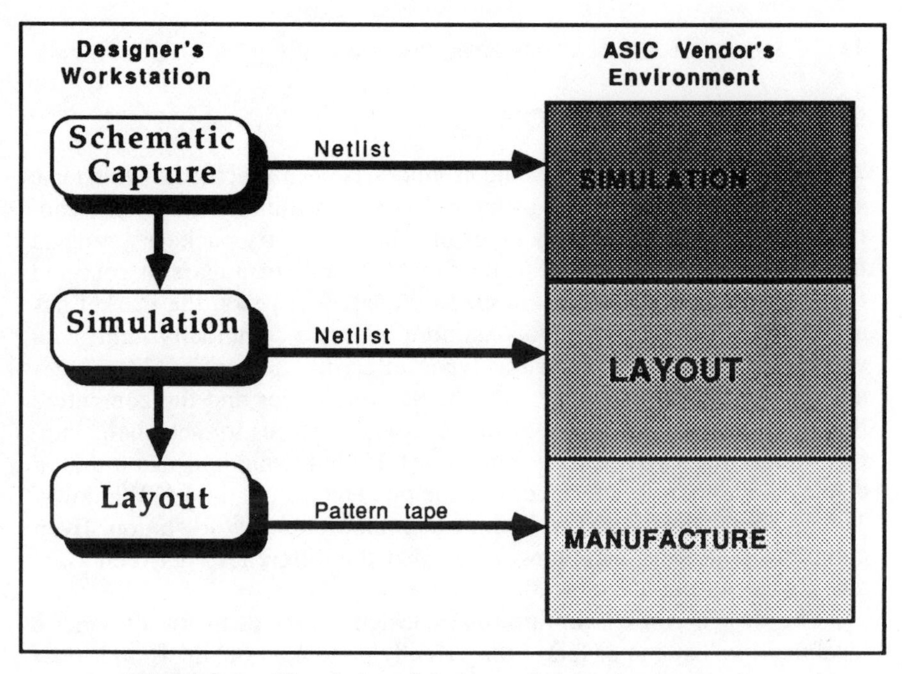

FIGURE 2-2. The choice of design paths a designer can take.

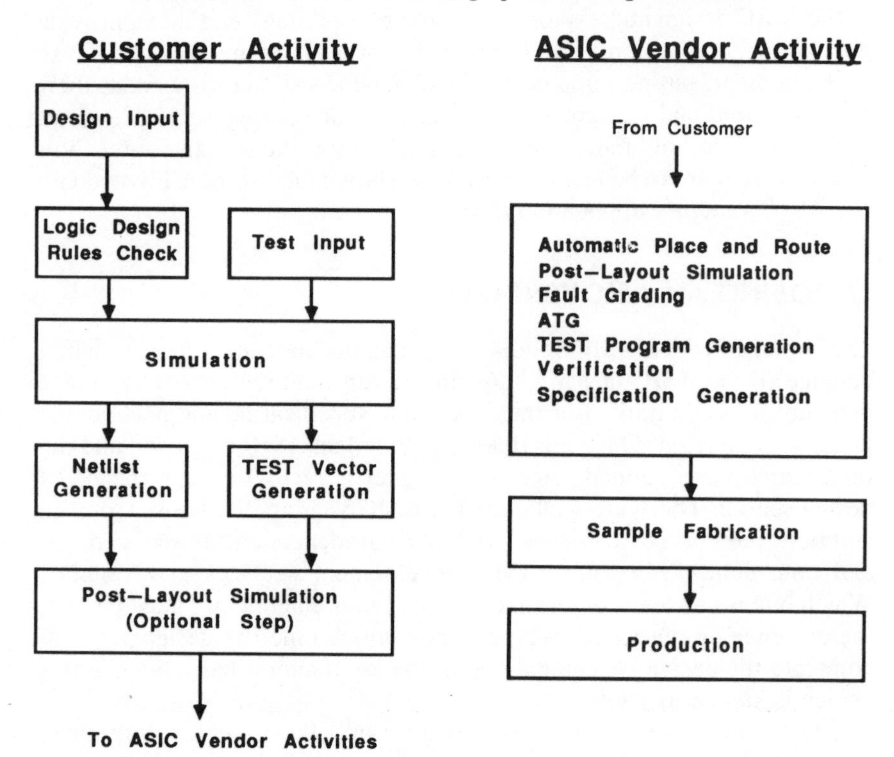

FIGURE 2-3. Typical designer and vendor responsibilities for an ASIC design.

DESIGN IMPLEMENTATION

The design stage begins, after the product has been specified, with a logic schematic of the new ASIC using an EDA workstation's schematic capture tool set. There are many kinds of schematic entry packages, ranging from a few hundred dollars to tens of thousands of dollars in cost and running on a PC or a workstation. In the last few years, the market has accepted schematic entry tools as more or less a commodity item, with very few differences among the types offered. The biggest differences among schematic entry types are the user interfaces and the computers that they run on. Most state-of-the-art user interfaces for schematic capture follow the OSF MOTIF standard for windowing and menu-driven displays. These user interfaces can include Open Look and X-Windows. The computer host is either a personal computer or a workstation. Even computer platforms are changing, so that the differences between a PC and a workstation are blurred.

The design can be captured at a design center or at the designer's local facility. The major factor that usually influences the decision is how much it will cost. However, the designer must look beyond the initial cost of the ASIC design and explore what advantages each vendor can provide versus the competition. The design center will allow the designer to use its equipment, but at some cost. If technical assistance is needed, there may be additional charges for applications engineering help. The costs will depend on how much work is being done at the design center, how many devices are to be manufactured, and how much work is involved for the ASIC vendor's application engineer.

CHOOSING AN ASIC VENDOR

The choice of which ASIC vendor to use can be, and most often is, tightly coupled to the ASIC design's implementation requirements. The choice need not be made until after the conceptual specification and partitioning phase is completed. Once it is determined that an ASIC is needed and the requirements are defined, then the designer is ready to begin the ASIC design search. There are well over 150 ASIC vendors to choose from. In addition, there are distributors, third-party (independent) design centers, and consultants who offer ASIC development services and support. Which one is chosen is determined by the requirements that must be met, the designer's experience level, and how much time the designer has to complete the design. A comparison of the key factors that influence this choice is shown in Table 2-2.

These factors show the commonality and the absolute differences among the choices. The base indicators can be misleading if taken at face

TABLE 2-2. ASIC Design Development and Support Choices

Service	*NRE*	*Technology*	*Methodology*	*EDA Source*	*Engineering Resource*	*Engineering Skills*	*Interface Flexibility*
In-house	Medium	All	All	High	High	High	No
ASIC vendor	Medium	All	All	Medium	High	High	Yes
Third party	High	All	All	Low	Medium	Medium	Yes
Distributor	Medium	CMOS	GA/PLD	Low	Medium	High	Yes

Source: ASIC Technology News, 1990.

value. For example, with an independent third-party design center or consultant, NRE charges are high compared with those for the other choices. However, the difference shows up in what you get for your investment. With an independent third-party design center, the designer gets more personalized attention and support, even if he or she is from a small company. The higher NRE charges up front often produce savings in the end because of the independent design center's skill and experience. All four services have the flexibility to let the designer do all or only parts of the design process.

After the choice of whom to use is made, the next step is to determine which is the best way to perform the actual design. The designer can use an ASIC manufacturer's design center, a distributor's design center, or an in-house design center. The designer can also choose to use a consultant to help with the basic design strategy and defining technology needs. A consultant can also be used as the interface between the ASIC design service center and the designer. For first-time ASIC designs, using an experienced consultant can actually reduce costs for the overall project. There are several third-party design centers that offer some of the same services as a consultant but on a broader scale. Third-party design centers offer assistance at almost any stage of the design process.

Vendor's Track Record

The user should take some time to research the history of the ASIC vendors that he or she has decided can meet his or her needs. The reason for this is to determine which ASIC vendor has the experience to back up its promises of success and support. One indication of a good ASIC vendor will be the existence of a second source for the libraries and technology.

Libraries

Most ASIC vendors provide a complete component data book showing their full library. The designer should take a good look at the data book before making a final choice of vendor to be sure that the information is clear and complete. Be sure to ask how much disk space a library will use. Disk space will be very important if the design is being done on the designer's own workstations. On some EDA networks, how and where the libraries are stored can cause performance, and hence productivity, degradation.

ASIC libraries consist of components that have been characterized by the ASIC vendor for both logic simulation and layout. These libraries can have different levels of functionality depending upon the application targeted by the ASIC vendor. For example, for a standard cell of a NAND gate, the user would have the ability to customize the timing on the pins to fit a specific application. Going one step further, for a standard cell consisting of a compiled ALU from a silicon compiler, the designer would need to specify critical parameters to fit the end application, then run the silicon compiler to generate the complete cell.

Libraries can include components that model anything from simple logic gates to microprocessor cores. All components have been characterized to be compatible with EDA simulators or the ASIC vendor's internal simulator. However, not all components in an ASIC vendor's library may be able to work with an external logic simulator. This limitation is especially true for those cells generated by a silicon compiler.

ASIC libraries must be purchased if they are to be used on a company's EDA workstations. The costs will vary depending on the type of contract arranged with the ASIC vendor and how many chips the vendor will be expected to manufacture.

ASIC DESIGN KITS

ASIC design kits specifying technologies and guidelines to be used with a particular EDA workstation vendor are normally provided by the ASIC vendor. A typical design kit is shown in Figure 2-4.

FIGURE 2-4. An NCR design kit for a Cadence Design system. *Courtesy of NCR Corporation, 1991.*

When investigating ASIC vendors, designers should ask each ASIC vendor the following questions and compare their answers before making a final choice.

Q. Which technologies does the design kit support?

A. All of the technologies the designer may want to use should be supported by the design kit for the EDA vendor's tools he or she will be using. It is not important to decide at first which technology is to be used, but the designer should have two or three choices.

Q. Who develops and maintains design kits?

A. Design kits should be developed and maintained by the ASIC vendor. This ensures full compatibility and guarantees design processing. The ASIC vendor is the only party who can effectively develop and maintain a design kit for a specific EDA vendor because the library is developed and maintained by the same ASIC vendor.

Q. Are the design kits guaranteed by the ASIC vendor?

A. Not all design kits will be guaranteed by the ASIC vendor for all EDA vendors. Be sure that the EDA vendor's tools are fully guaranteed. If not, problems after fabrication could be very costly.

Q. Does the EDA vendor qualify the design kit?

A. Quality-conscious EDA vendors will qualify the design kits for their systems through a design kit review program. Design kit review and qualification offers both parties beneficial checks against problems that could cause the designer unnecessary trouble. The ASIC vendor normally pays a fee to have its design kits qualified by the EDA vendor.

Q. What assurances are there that the design kit will be compatible with new EDA products and software releases?

A. ASIC vendors that participate in regular review and qualification programs with their supported EDA vendors are assured of 100-percent library compatibility. Normally, the EDA vendor will supply early copies of new software tools several months prior to release to assist with testing of libraries and modifications to design kits.

Q. Are the ASIC library models compatible with other parts models?

A. All model types supplied by the EDA vendor should be compatible with the ASIC library to ensure full design testability when the ASIC is connected into the target board design. These models include standard parts, other ASIC libraries, and other third-party libraries supported by the EDA vendor. Vendors should offer migration paths from a gate array netlist directly into a standard cell implementation.

Q. Are the same libraries used for functional simulation, timing analysis, and fault analysis?

A. Normally, one set of libraries is all that is required to support the entire range of analysis tools from the EDA vendor. Be wary when different model libraries are used for different analysis tools, since it is very difficult to guarantee results from phase to phase. If multiple model sets are required, the costs may be more than the new design can bear, with reworked silicon the result.

Q. Is technology-independent ASIC design supported?

A. Having technology design independence provides for generic schematic capture and simulation. This, in turn, provides the designer with the ability to perform extensive, but less costly, "what if" analysis on the new design without committing to a specific technology.

Each design kit contains a manual with a step-by-step procedure organized to lead a designer through the ASIC design process. These process steps hold true whether the designer is using the ASIC vendor's facilities or his or her own EDA system, as long as he or she is using the ASIC vendor's libraries. The ASIC vendor design kit contains a design kit user's guide, the vendor's libraries (normally on magnetic media), and vendor application programs (normally on magnetic media).

Normally, if the designer is using a kit for a specific EDA vendor, the applications programs supplied with the kit will contain enhanced applications based on the EDA vendor's design tools. For example, if the designer was designing on a Mentor Graphics IDEA station, the user would use an enhanced NetEd to capture the schematic. Symbols would be taken from the ASIC vendor's macro libraries, then placed on schematic sheets. The designer would use the ASIC vendor's enhanced version of Mentor Graphics' Symed to create new, user-defined symbols and macros.

After the schematic is captured, the designer must compile the design to prepare it for design rule checks (DRC) and simulation. The ASIC vendor supplies special application programs that analyze the circuit to verify compliance with the vendor's design rules. Normally, this DRC program provides error reports and some kind of cell utilization listing to guide the designer through the presimulation process.

Another special program supplied by the ASIC vendor determines delays and inserts them into the compiled database for postlayout simulation with real layout values.

ASIC VENDOR GUIDELINES, ASSISTANCE, AND SUPPORT

ASIC vendors provide design guidelines as part of their design kits. These guidelines provide the optimum ways to use the vendor library so as to avoid costly errors. Additionally, ASIC vendors have application engineers who can assist the designer from the design entry step through the layout and verification process. Most of the AE support, however, will be provided at the design center and may cost extra, depending on the level of expertise.

ASIC vendors provide application notes to assist designers in using the vendor's design kit, especially when using new technologies. In addition, most ASIC vendors provide designers with in-depth training sessions that cover all aspects of the type of design being performed.

DESIGN IN-HOUSE, RENT, OR USE THE ASIC VENDOR DESIGN CENTER?

The biggest question that confronts smaller companies or new departments doing their first ASIC is whether to buy workstations, rent or lease workstations, or just use the ASIC vendor's design center workstations.

The cost of purchasing a full-function EDA system is nontrivial. At a minimum, the system must be able to perform design capture and some simulation. However, depending on how the design is captured, you may have to pay to have the design simulated on the ASIC vendor's simulator. Further, depending on what the ASIC vendor uses for a simulation engine, you may be paying for time on a mainframe. The computer costs can be quite expensive, so you should check the ASIC vendor criteria for acceptance of EDA vendor simulation inputs to avoid resimulation charges. The postlayout timing analysis step can be performed by the designer, or the ASIC vendor can do it. If the ASIC vendor's engineers perform the postlayout timing analysis, there will be an additional NRE charge. However, having the ASIC vendor perform the simulation can be the best option if the size and complexity of the design warrants. If the design is large (greater than 25,000 primitives), it may be more cost-effective to have the ASIC vendor use its computer than to perform the simulation on a workstation. The exception to this is if the designer owns a hardware accelerator.

POSTDESIGN ACTIVITIES: BACKANNOTATION FOR SIMULATION

During the design stage, the ASIC is simulated using estimated delay calculations. These delays are based on equations provided with each of the library components supplied by the ASIC vendor. After the ASIC is ready for layout, actual capacitive loading parameters are calculated and "annotated" into the original schematic and simulation database. The design is then resimulated to make certain that the real delays will still allow the design to operate within the specified limits and to perform the functions it was designed to perform. More discussion of this area has been deferred until Chapter 7.

SUMMARY AND CONCLUSIONS

The ASIC design process can vary among ASIC vendors. The process flow, however, is relatively similar among ASIC vendors and companies that develop their ASICs in-house.

There are several questions that must be asked and answered when going through the process of choosing an ASIC vendor. Each new design may require different considerations that will affect the choice of a single ASIC vendor. If requirements for multiple new designs are understood, a large cost savings in NRE and library charges can be realized. Knowing the ASIC vendor's capabilities will prevent problems at times when the designer cannot afford them (Murphy's law).

Saving money is usually a major objective of ASIC design. It pays to understand an ASIC vendor's standard services, what costs are extra and how much they are, and what prototypes will cost. The designer must work with the ASIC vendor to determine which type of approach will be most cost-effective: gate array, standard cell, or full custom.

First-time ASIC designers who do not own EDA tools can cut costs by weighing NRE costs against the cost of buying or leasing EDA equipment for that first project. It may also be cheaper to perform the entire design at the ASIC design center or to use an independent design center.

REFERENCES

Friedman, M., Selecting the Right Supplier for Mixed-Signal ASICs. Sierra Semiconductor, *Computer Design* April 1990.

Gifford, M. High Density Designs Require a New Set of Rules for ASIC Engineers. *ASIC Technology & News,* March 1990.

Homstad, Gerald E. and Robert B. Smith, How to Deal with a Vendor. Gould AMI, *Electronic Products,* November 1989.

Kahle, Charles A., Producing Complex ASICs on Tight Schedules. Unisys Corp, *Semicustom Design Guide,* 1989.

LSI Logic Corporation. 1989. *LSI Logic Primer for ASIC Design.* Milpitas, CA, LSI Logic Corporation.

Motorola Inc. 1989. *Motorola H-Series Design Guidebook.*

NCR Corporation. 1989. *NCR ASIC Data Book.* Dayton, Ohio.

OKI Semiconductor Inc. 1989. *OKI Design Guidebook for Mentor Graphics Workstations.* OKI Semiconductor Inc.

T.I. Corporation. 1990.

VSLI Technology, Inc. 1989. *VLSI Design Guidebook for Mentor Graphics.* VLSI Technology Inc.

3

Design Creation for ASICs

INTRODUCTION

Design creation is the next step in ASIC design after concept design and specification definitions have been completed. ASIC vendors refer to this stage as the design implementation phase. Design creation for an ASIC is not just schematic capture, but an iterative, hierarchical implementation process including both design *and* analysis steps. The complexity of new ASICs with more than 100,000 gates brings new design challenges because of the detailed information that a designer must understand and deal with. Design details at the structural (gate) level for a 200,000-gate ASIC require new design techniques that help the designer maintain control of these complex designs. New-generation design techniques use the latest technological developments offered through electronic design automation to support complex ASIC design.

SYSTEM ON A CHIP

Today's ASIC designs should be approached like a ''system on a chip'' by choosing and connecting functional cells from an ASIC library. This approach allows designers to create and verify a new design's specifications during the concept and partitioning phases at the architectural level (see Figure 3-1). At the architectural level, a designer can perform trade-off

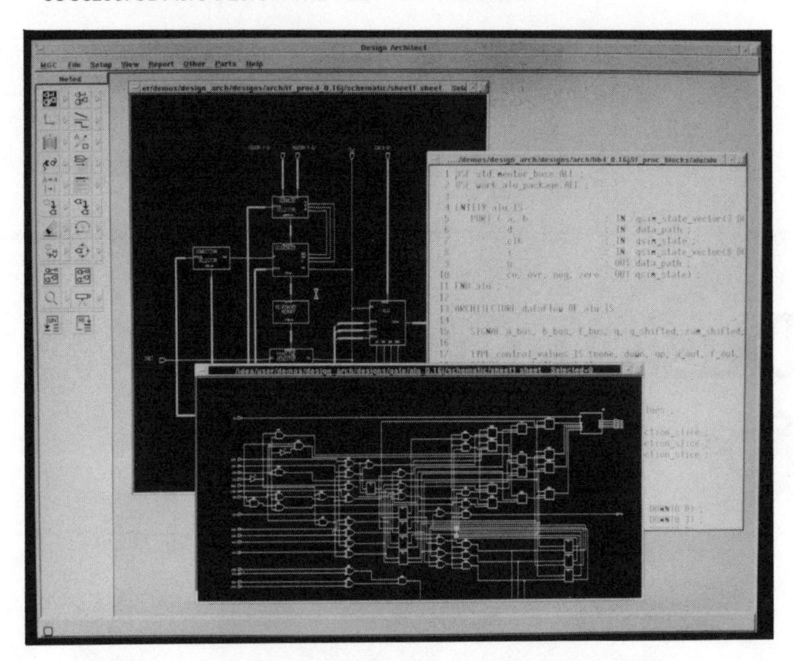

FIGURE 3-1. Design creation example with design architect.

analysis to determine top-down specifications *and* optimize the design. EDA tools make this possible through high-level behavioral constructs, which can be decomposed into functional gates or macro-level functional blocks. This concept introduces design flexibility since the designer can make changes at several levels across the ASIC's functional block elements and see the results. Major parts of the design can be left at the behavioral level while a specific block is "synthesized" as gates to perform detailed verification. Each major segment of the ASIC can be treated the same way, with full functional checks at intermediate levels across design partitions. The ability to deal with complex designs at a high level improves the designer's productivity by keeping the design and its specifications in perspective at all times. This saves time in new ASIC design by increasing simulation efficiency and increases a single designer's or a design team's productivity.

This chapter will focus on creation of a new ASIC design using EDA tools (see Figure 3-1) and will discuss the use and development of libraries and models, both the vendor's and user-defined. This section will also discuss the uses of the VHSIC Hardware Description language (VHDL) and how it is being applied to ASIC design. Later, in Chapter 4, the process of logic synthesis and the step-by-step decomposition of an ASIC design from a high-level behavioral description to a structured description is described.

START THE DESIGN AT THE ARCHITECTURAL LEVEL, FROM THE TOP

A new ASIC can begin several ways. With the increase in ASIC complexity, which is expected to continue, designs can be successfully created only by starting at a conceptual or functional-block level (see Figure 3-2). The successful ASIC designer works with high-level behavioral blocks to develop major functional sections that support the ASIC's specifications. These behavioral blocks can include everything from basic registers/latches to microprocessors and memory. Trade-offs for performance, cost, and functionality are made using behavioral models. Design details such as number of registers, datapaths, and bus widths can be identified. Then the design can be partitioned and each major subsection designed in complete detail at the structural (gate and transistor switch) level. This method allows the designer to iterate across the whole design in segments

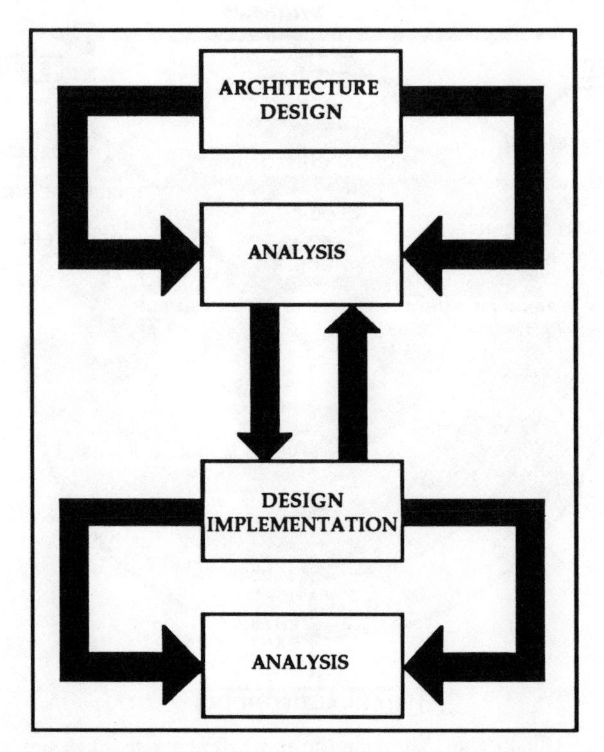

FIGURE 3-2. Multilevel design, partitioning, and verification. *Courtesy of Mentor Graphics, 1990.*

using decomposition techniques, usually with logic synthesis. The design process at this level is usually segmented into two major steps: functional or detailed design and verification. Details of each step will be described in the next several sections. This process should be coordinated as part of the overall system design process, and these designs should be proceeding concurrently.

Looking at the ASIC design process using the Y chart (Gajski and Kuhn 1983) from a functional level (Figure 3-3) allows one to segment the design process into three basic process groups:

- Behavioral: high-level constructs, language-based
- Structural: low-level logic, precise constructs
- Physical/geometry: layout constructs

Each process area uses several levels of top-down functionality for the design process. The primary starting place is a behavioral description

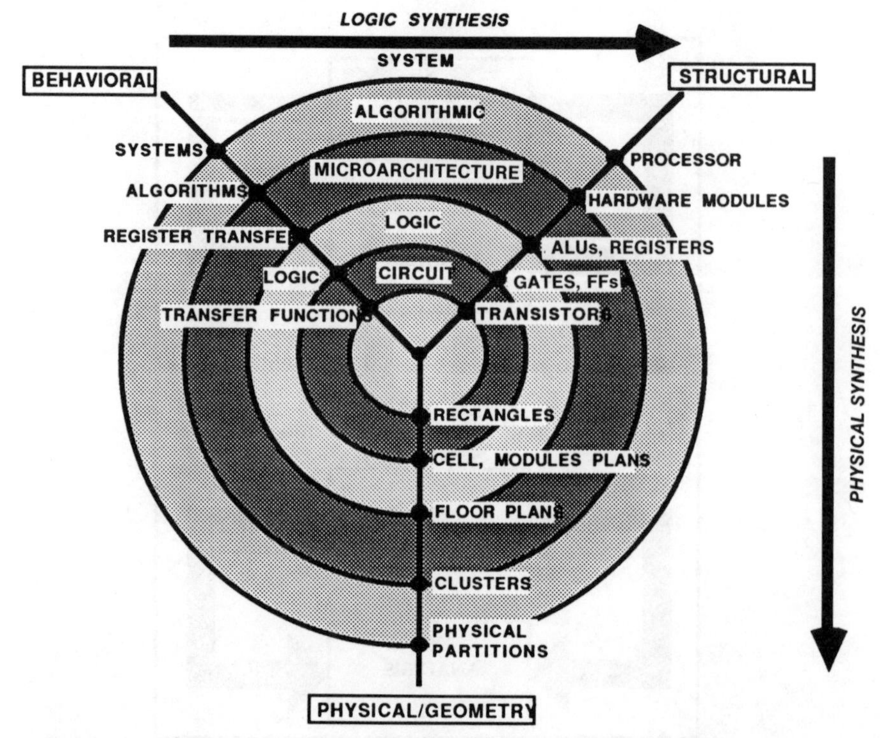

FIGURE 3-3. Gajski chart showing the three dimensions of design description: behavioral, structural, and physical.

of the circuit, including inputs and outputs. The next logical progression is to decompose the high-level constructs into more precise functional units operating more similarly to the way the actual hardware will perform. The last step in the progression is to transfer the precise functional units into physical elements that can be manufactured.

A question that usually comes up is, is there any way to bypass the structured phase and go right to physical layout? Silicon compilers offer a possible solution at this stage. Silicon compilers allow a high-level behavioral description, usually from a special language, to be compiled directly into physical geometries for layout. But how well do they work? Later in this chapter silicon compilers will be described, and how they can fit into the ASIC design cycle will be discussed.

BEHAVIORAL-LEVEL DESIGN

The behavioral level is used to describe a new ASIC as a system including inputs and outputs. VHDL models are usually used as the behavioral description since they are more flexible and can be used at various levels (see Figure 3-3, Gajski chart) from basic transfer functions to algorithmic or architectural constructs. The objective at this level of design is to ignore a lot of detail and concentrate on specifying what the design will do at a major block level. The benefits to the designer are that this provides a fast and accurate way to verify a design at an intermediate level and offers a faster and more efficient means of "what if" analysis.

From the chart, one can see that decomposition from a behavioral to a structural level can take place on several planes. Behavioral logic functions can be synthesized along the logic plane to create gates or flip-flops. The structural level can be further synthesized for layout to account for physical device creation.

VHDL: A Great Place to Begin

Today's designer must have a working knowledge of the VHSIC hardware description language (VHDL). VHSIC is the Department of Defense acronym for very-high-speed integrated circuit. VHDL is rapidly becoming the industry standard language for developing behavioral descriptions of ASIC designs. These descriptions can then be compiled and simulated using most standard EDA simulators.

VHDL is based on the Department of Defense's efforts with hardware description languages and design development. Its goal is to provide a standard specification language for defining a simulation model that can obtain the same results on different simulators. VHDL was adapted as a standard design language by the IEEE and named IEEE 1076-1987. Formal standardization was encouraged when the Department of Defense mandated that all ASIC designs must be traceable to the IEEE-1076 standard or they cannot be used in government-related designs. (See the further discussion in Chapter 4.)

VHDL is used to develop high-level functional blocks that can be decomposed using synthesis techniques or manually as design verification progresses. VHDL can be entered either as a language code or through the use of graphical constructs to create behavioral descriptions of circuit modules.

Designing with VHDL

The 1980s saw designers looking for help in designing complex systems with over 500,000 gates. Behavioral language modeling methods were

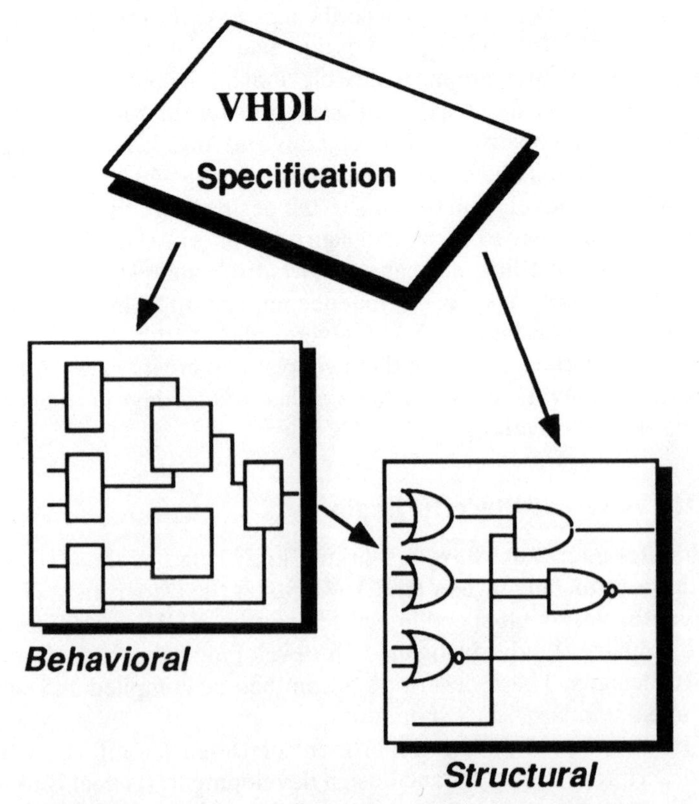

FIGURE 3-4. VHDL can be used for high-level behavioral design and for detailed structural design.

employed to try to solve the problem. But behavioral language models put a strain on hardware design engineers, who had to learn programming techniques. Hardware designers wanted a modeling technique that more closely tracked hardware design, and they finally got it with VHDL (see Figure 3-4).

After architectural, behavioral, and functional design checkout are completed, the design is ready for decomposition into structured gates. This entails translating behavioral circuit descriptions into structural representations for schematic creation.

FUNCTIONAL DESIGN

Functional design deals primarily with the development and verification of the logic executions (and their interactions) of the ASIC's high-level blocks. The functional level allows a designer to verify specifications at a level where the details can be dealt with, and still verify the entire ASIC. To perform functional design, each of the major blocks must be specified, developed, and verified (see Figure 3-5). The process for this design method is to build each of the blocks in the proper development sequence for design and verification. The proper sequence might mean building these high-level blocks using a serial or parallel approach. The goal is to make certain that each high-level block is verified against its specifications relative to other blocks that it will interact with.

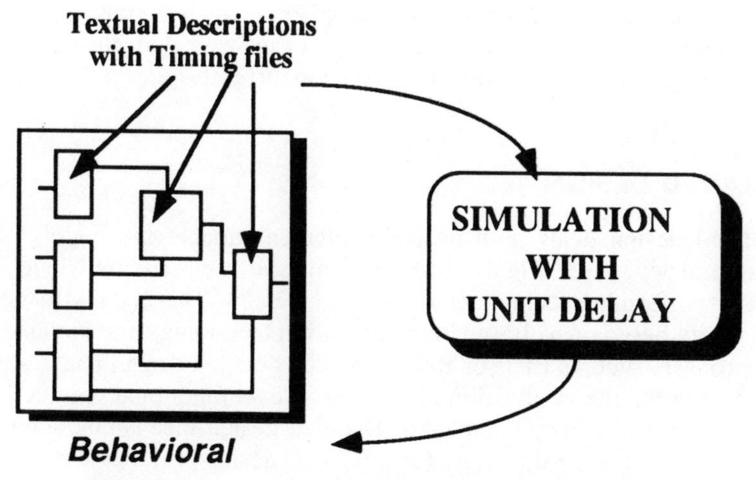

FIGURE 3-5. Functional design and simulation steps.

The best way to begin is by developing a detailed VHDL, IEEE-1076 description of each functional block. These functional blocks can then be matched to an ASIC vendor's cell library and a detailed design developed.

STRUCTURAL DESIGN

Design at the structural level begins to approximate the way the actual hardware will work. This phase includes all the details for the complete design, including component level schematics and the necessary information required for analysis.

The structural stage can be frustrating if the designer takes on the detailed implementation of a 50,000-gate design. Partitioning the design into high-level and low-level segments makes the structural phase smoother and less complex. For example, take a 250,000-gate ASIC design with about 100 gates per sheet. To develop this design at the structural level would require about 2500 sheets! Too much detail. No one designer can cope with all the logic and timing details for a design with this amount of complexity. There are other ways to tackle structural development of complex ASICs.

The development of the structural segment can be easily done using logic synthesis techniques, removing the frustration of dealing with the detailed complexity. Logic synthesis tools provide a schematic representation of structured gates to be synthesized from high-level VHDL or behavioral circuit descriptions. Today's EDA design tools allow ASIC designers to work concurrently at multiple levels. These same tools allow designers to be more efficient even when dealing with new methodologies. The design creation tools from EDA suppliers provide hierarchical design entry and analysis tools, allowing designers to work with the behavioral level and gates (or transistor switches) concurrently.

DETAILED DESIGN

Detailed design deals with design implementation at the logic gate or structural level (see Figure 3-6). Designing at the structural level also means performing technology-dependent design. This means to design and verify based on a chosen implementation technology and manufacturing process, such as ECL or BiCMOS. The detailed design phase should not begin until the verification of the higher-level functional specifications has been satisfactorily completed. Detailed design means characterizing all of the low-level parameters for the ASIC design at the gate or transistor level.

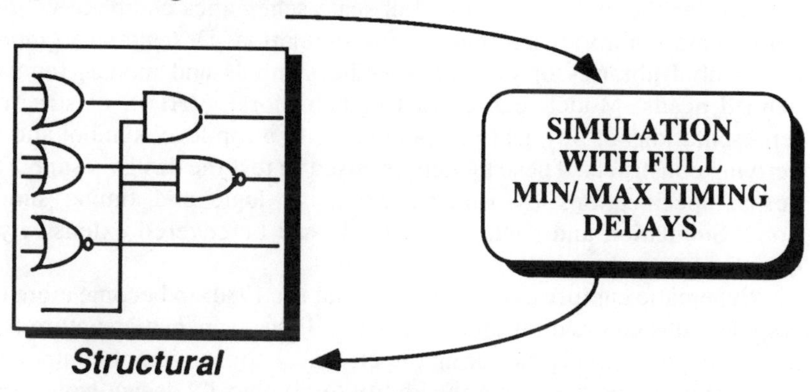

FIGURE 3-6. Detailed structural design and simulation steps.

Specifying the details for loading and timing characteristics is a complex but rewarding task. It is complex because every critical data path, every I/O pad, and every clock path must be calculated and measured against the specifications. ASIC vendor data books describe each of the building blocks (cells) and their timing equations. The rewarding aspect of specifying all the timing is that it saves many headaches later in the design cycle. A designer can quickly be overwhelmed at the detail level for a 100,000-gate ASIC.

PHYSICAL/GEOMETRY

The next phase in the ASIC design is to develop physical descriptions from the completed structural design to create a manufacturing database.

This phase of the design process is normally handled by the ASIC vendor at its facilities with little participation by the ASIC designer. The designer's portion of the layout step is complete when the pattern layout tapes are delivered to the ASIC vendor.

"STANDARD" ELECTRONIC DESIGN AUTOMATION AND SCHEMATIC CAPTURE

Schematic capture is the grass roots of EDA. Originally known as electronic drafting, schematic capture tools allow designers to develop electronic designs without pencil and paper. The most important aspect of design creation tools is their integration with EDA analysis tools. All major EDA vendors provide schematic capture tools for design creation. Many offer the ability to design full-scale schematics complete with generic symbol libraries and models for simulation. Designers can modify the symbol libraries or develop specific symbols and models for user-defined needs. Models can consist of behavioral, VHDL, or structural representations of physical components with a top-level symbol and underlying constructs. These models are used to test the design's support of design specifications by running extensive logic and timing simulations. Simulation and simulation models will be covered extensively in Chapter 5.

Schematic capture evolved throughout the 1980s to become more of a complete design creation tool. In the late 1980s, schematic capture tools moved beyond simply capturing the electrical connection of components in a design by incorporating the ability to graphically design higher-level component constructs. By giving designers the ability to do more than just draw curves and polygons, these tools give them more flexibility and enable them to realize increased productivity by stimulating their creativity. The decade of the 1990s should see design creation tools truly provide a complete approach to systems design, including software, hardware, and mechanical design.

What to Look For in a Schematic Capture Tool

- **General design considerations—EDA vendor.** Most EDA vendors provide a proprietary schematic capture package for design creation. Most of these same vendors allow the designer to use behavioral- and structural-level components in the design. For ASIC design, the designer will begin by using high-level blocks and then decompose down the hierarchy to structural cells. Today's standard design creation packages offer full mixed-level design support.
- **Determine minimum expectations.** Minimum expectations should be centered around the user interface of the tool: how easy it is to prepare a design for simulation, and how easy it is to make changes in the schematic (major and minor) and be ready to resimulate. The time for making a change and getting back to simulation is commonly known as design iteration time, or DIT.
- **Use the designer's schematic capture tools or the ASIC vendor's?** If the designer already owns a design creation package and has used it for several designs, then he or she should plan on design capture using this system. The benefits of this are that the designer can use any macros, special symbols, or other company-specific capabilities that were useful on previous design projects and that the learning

curve is reduced. The designer will have to load the ASIC vendor's cell libraries onto his or her network or workstation. Using his or her own design creation tools means that the designer will have to provide a complete schematic netlist to the ASIC vendor for continued processing.

A designer who does not own a design creation package should consider using the ASIC vendor's tools or a design center. This is especially true when there is a critical time-to-market window. A designer cannot afford to be evaluating EDA tools and trying to learn a new EDA toolset while trying to get a high-quality product to market quickly.

An important item to check on is the NRE costs associated with the ASIC vendor's engineering support. Whichever method is chosen, the NRE can play an important role in overbudget design costs. Make certain that it is understood exactly which services are free and which cost extra.

ASIC Libraries and Simulation Models

ASICs are designed using libraries of symbols whose underlying functionality is understood by a simulator. Symbols can represent anything from a transistor switch to a complete microprocessor. ASIC libraries are supplied by the ASIC vendor as part of the ASIC design kit. These libraries contain the cells that provide the symbols and functions used for schematic capture and simulation. An ASIC library cell is described by a symbol, a function table, and a description of specific parameters for the cell. Figure 3-7 shows how a typical ASIC cell model is specified. Nor-

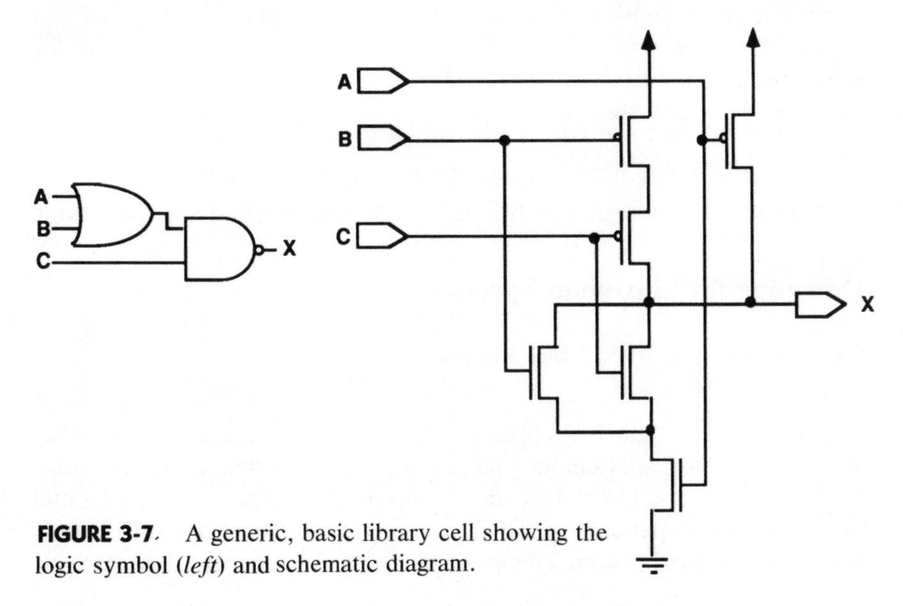

FIGURE 3-7. A generic, basic library cell showing the logic symbol (*left*) and schematic diagram.

mally, the designer has to add only expected timing parameters, signal names, and pin names. The design process is to select symbols from the ASIC cell library and enter them on a schematic sheet. The designer then makes net connections between the pins on the symbols. All pins are given names, and timing parameters are defined for each. Timing parameters can be a single number or an equation.

SPECIFIC DESIGN CONSIDERATIONS

A primary goal of the new ASIC designer is to provide a design that is optimized for power, speed, die area, and package type. This new ASIC must also fulfill the design specifications that describe the full functionality that the ASIC must be capable of providing. Let's take a look at what some of the parameters of interest are.

Designing for Die Size Optimization

- Check ASIC vendor requirements or recommendations
- I/O placements and number of pins needed
- Preplacement of cells and floorplanning

Optimizing for die size means using the smallest amount of silicon area for a given design size (number of gates). Taking the time to determine optimum die size will save development time and ASIC production costs. The ASIC vendor's documentation and an application engineer can help guide a new designer to achieve optimal die size. The benefits for achieving optimal die size are:

- Smallest gate array package—reduces costs
- Helps achieve higher-speed designs—maximum performance
- Helps meet power constraints specifications—increases reliability

Designing for Maximum Speed

Use NAND logic over NOR logic wherever possible

Remember EE design 101? This was where you learned about the mobility of electrons and that N-channel transistors are faster than P-channel transistors. This is due to the fact that in a NAND configuration, P-channel transistors are in parallel, resulting in more symmetrical transitions than in the NOR configuration. The symmetrical transitions translate into improved speed through lower propagation delays in the circuit. Additionally, the channel resistance of an N-channel transistor is almost half that of its P-channel counterpart. These facts indicate that designing for speed means using the right logic functions.

Avoid extra delays

When doing floorplanning for an ASIC, minimizing the effects of interconnect delays is imperative for maximizing ASIC speed. The interconnect delays can be optimized by proper partitioning of the ASIC logic. This is especially important when dealing with ASIC technologies in the 2-μm range and below. Much of this can be avoided by trying to minimize wire lengths.

Fanout and loading considerations

- **Using high-drive buffers.** To avoid the effects of parasitic capacitance when using buffers, use the smallest internal buffer that meets the needs of the design. Excessively high-drive buffers will present a large load to the driving cell and will increase capacitive loading, slowing down the circuit.
- **Use buffers for balancing fanout loading.** Using buffers strategically throughout the circuit will help eliminate fanout loading problems. Each ASIC vendor will have recommendations on which cells to use as buffers for different loading applications. "Loading applications" refers to the number of loads to be driven as well as the driving cell.
- **Using AND-OR-INVERT.** Many ASIC vendors recommend the use of AOI cell configurations based on loading factors. The primary reason is that the AOI configuration performs two levels of gating with only a single inversion. AOI cells are provided with different variations of drive and propagation delay characteristics.
- **Avoid excessive fanout.** To avoid speed problems, reliability problems, and testability problems in an ASIC design, pay close attention to excessive fanouts. Most ASIC vendors provide guidelines for fanout loading for configured cells. When using macrocells, be sure to include all internal loads in the calculations to avoid hidden problems. Internal cell loads are not always visible in the descriptions provided. Ask for schematics of all macrocells with internal cells identified to determine actual fanout and loading.

Improving delay times

There are several ways to improve circuit delay times for ASIC designs. To maximize the efficiency of the ASIC, a designer must understand the trade-offs that can be made using the chosen ASIC vendor's cell libraries. ASIC vendors provide application engineering assistance plus design guidelines to help the designer achieve the desired results.

- **Suggestions for using counter cells.** When using counter functions in an ASIC, it is recommended that the use of shifting counters rather than binary counters be considered. A Johnson counter is an example of a shifting counter. A Johnson counter is faster than a binary counter because no gating is required between flip-flop stages. The downside to using a Johnson counter is that achieving the proper count sequence may require extra flip-flops. Therefore the designer must decide which is more important, speed or gate count.
- **When to use predefined macro-level blocks.** Where speed and design time are critical, ASIC vendors recommend designing with predesigned macro-level blocks. The trade-off for complex functions will be speed versus gate count. The predesigned macro-level function block will probably require more gates but will normally save several nanoseconds in critical delay paths.
- **Speeding up critical path signals.** Several ASIC vendors suggest using additional gates when trying to speed up a signal in a critical path. By using duplicate logic gates, a designer can reduce the fanout and gain speed for the critical path. An example of this technique would be an input gate with a high fanout where one of the signals is in the critical path. By duplicating the input gate and running the critical output through a single gate, the signal is sped up and the fanout for the other gate is reduced.

Design Techniques Used for Increasing ASIC Reliability

This section discusses how to ensure that a new ASIC design will be reliable after it is developed. Designing for reliability will reduce development costs and ensure a smooth test cycle for the ASIC. In most cases, the ASIC vendor's application engineers will have a list of criteria similar to the ones here and will work with the designer to meet the checklist.

Synchronous design techniques

Most ASICs are designed as synchronous devices. ASIC vendors will certainly recommend the use of totally synchronous designs because process variations may cause unpredictability or spike generation. These problems make the test cycle more expensive and time-consuming and will lead to redesign of sections of the ASIC to correct them.

Eliminate glitch generation

That's just good design practice. In ASIC design, because of the circuit speeds it is even more important that the signal be timed properly. Unexpected spikes in a circuit can cause improper clocking or initialization of critical portions of the design. Problems like these cause delays in development and contribute to costs.

- **Gated clocks are not recommended.** The use of gated clocks in an ASIC design can cause false clocking of flip-flops and latches. The false clocking causes the wrong data to be latched, which leads to improper operation. Timing tolerances in an ASIC are critical, leaving little margin for process variations. ASIC vendors recommend that gated clock designs be avoided. The result will be a much cleaner and more reliable design.
- **Eliminate race conditions.** More EE101. Pay close attention to clock signals that may split and then reconverge at a common gate input (known as reconvergent fanout). Race conditions and hazards should be avoided at all costs because of the unpredictable nature of the signals involved.

Don't leave floating nodes

Three-state buses are a source of concern because of the extra logic state of "floating." Floating nodes can generate oscillations or noise if used on an internal input node. Generally, using three-state buses improperly can cause bus contention problems. To ensure proper operation when using three-state internal buses, use an extra buffer gate to drive the bus when its signal inputs would cause a floating condition. Most logic simulators can detect bus contention problems, and can provide a report that tells the designer which input is causing the error. Some ASIC vendors do not allow any bus contention to exist.

Transmission gates must be buffered

Transmission gates can provide high speed for light load applications but can also bring reliability problems. Transmission gates have a signal degradation problem based on the P-channel and N-channel configuration and its operating characteristics. The nature of these devices requires both the P-channel and the N-channel transistor to accurately transmit both a logic 1 and a logic 0 without degradation. Designers using multiplexer or storage cells made from transmission gates should ensure that there are buffered outputs to provide isolation and to eliminate signal degradation. Some vendors provide output buffers for their latches and flip-flops to ensure reliability of the design. If using vendor cells, check to make sure the buffered outputs are included. If using your own devices, adding buffers is advised to avoid problems. Also, designers should be careful of false triggering in feedback paths with transmission gates as a result of poor locking of the gates. Designers should make sure that their design will properly hold the transistor gates in a fully off position to remove the possibility of false triggering.

Buses must be controlled

ASIC vendors recommend paying special attention when using buses in an ASIC design. Poor bus routing can produce parasitic capacitances, routing problems, poor floorplans, inefficient chip utilization, and incorrect die-size estimates. Most of the problems result from the use of lower-level buses that run to several places in the design. These buses are buried in the lower levels of the design hierarchy; they are not easily seen and therefore cannot be planned properly. As much as possible, keep buses visible at the top level of the schematic to aid in the proper planning of bus routing. Buses that connect sublevel blocks are okay and provide tight coupling between the blocks. Good floorplanning can take care of proper placement to avoid any of the problems associated with buses. Also, bus routing with a gate array can cause problems because of the limited number of routing channels. In the case of a gate array, parasitic loading can be a major concern.

Don't guess with I/O pins

All ASIC vendors provide stringent guidelines for setting up I/O pads on the chip. Strict adherence to these guidelines will ensure a reliable, successful ASIC. I/O guidelines include power and ground as well as signal inputs and outputs. I/O signals fall into four categories: power and ground, inputs, outputs, and bidirectionals.

All inputs are buffered and are CMOS- and TTL-compatible. All inputs are routed through an input pad and a buffer; the input pad protects the circuit from excessive input voltages and electrostatic discharge. Inputs can be used with or without pull-up or pull-down resistor configurations. Most ASIC vendors provide special Schmitt trigger cells that allow input signals to be set to switch at voltage thresholds other than TTL logic levels. Check the ASIC vendor's data book for the exact specifications for a Schmitt trigger input pad that will meet the requirements of the design.

Output signals must be connected to the outside through output drive pads and must be carefully placed because of the effects they can have on clock and control signal lines. Output buffers come in different types—inverting or noninverting, with or without pull-up or pull-down resistors, and even a tri-state configuration. Output pads offer different drive current ratings, with a standard pad at 2–4 mA and a high drive up to 48 mA. High-drive output pad cells are normally available only for standard cell designs. If a gate array is being used, the most common technique is to use adjacent output buffers in parallel to gain additional current drive. Normally, there is a limit of six buffers (@ 4 mA each) that can be used in parallel to provide ~24 mA of drive current. Check with the ASIC vendor's application engineers to determine the exact limitations for the design.

- **Power pads and drive current.** Standard output buffers provide 4 mA of drive current. Most ASIC design standards, for CMOS and TTL, have a limit of one V_{SS} pad and one ground pad supplying up to sixteen output buffers. This limit is normally for pins switching simultaneously and driving into 50-pF loads. By following this guideline, the number of power pins that an ASIC will require can be determined.

  Optional outputs can be used to drive from 2 to 48 mA. With the parallel output buffers technique a chip can supply additional drive as needed. However, if the design requires more than the standard output drivers or if it is necessary to parallel output buffers, then the designer should consult with the ASIC vendor to determine the best method for properly defining the chip's power pin requirements.

  It is recommended that power pad placements occur in an orderly fashion to properly support the output buffers' switching needs. For example, NCR requires the first V_{SS} pad to be located on the opposite side of the chip from the first V_{DD} pad. As additional power pad pairs are required, they should be located close to the output pads.

  Power pads tie into metallized planes in the chip with a separate plane for power and ground. All power connections are tied together through this plane. These power planes are designed into the chip to control or reduce signal I/O path inductance, V_{DD}/V_{SS} path inductance and resistance, and signal-to-signal capacitance.

  Additional attention should be paid to understanding pad placements and the type of packaging. Packaging types have different pad placement restrictions. For example, pads are not allowed to be placed in die corners for ASICs that will be packaged in plastic.

  The designer should refer to the ASIC vendor's specification sheets for package types and pad placement restrictions.

- **Eliminate CMOS latch-up.** Latch-up of the device is a standard problem with CMOS technology, and if the designer is not aware of the potential hazards, it can cause the device to be destroyed. There are a few important rules that a designer should follow with CMOS ASIC designs to avoid the possibility of latch-up.

  1. Be sure to have current limiting to the supply current source. Additionally, ensure that the inputs and outputs are limited to their maximum *rated* values.
  2. Add external protection diodes to sensitive inputs and outputs where there is a possibility of voltage transients that could cause device latch-up.
  3. Power-supply runup should be sequenced to ensure that all the power-supply pins are at full power *before* any inputs or outputs become activated.
  4. Try to avoid unnecessary or excess noise on power-supply lines by providing voltage regulators or filters in the final board design.

- **Clock and control signal placements.** Never place clocks or major control signals, such as reset, between V_{SS} and high-drive output buffers. The safest recommended position for clock pins or major control pins is between two V_{SS} pads, although a clock pin can be positioned with a V_{SS} pad between it and a high-drive output. This placement will minimize switching noise and improve the reliability of the design.
- **Bidirectional pins need pull-up and pull-down resistors.** Increased noise margins can be achieved by using pull-up or pull-down resistors with bidirectional output buffers. ASIC vendor libraries provide bidirectional pads that are combined with an input buffer and an input/output buffer. To increase the noise margin when using this configuration, vendors also provide bidirectional pads with pull-up or pull-down resistors. There are several configurations of the pull-up or pull-down resistors that allow the designer to specify the amount of output drive current desired. ASIC vendors do not recommend the use of pull-up or pull-down resistors for setting up floating signals. Use of an off-chip high-precision resistor for setting correct levels on floating inputs is recommended.
- **ESD and latch-up.** Normally, latch-up will not be a problem with CMOS; however, it is a good idea to be aware of the causes and some methods of prevention to ensure clean operation of an ASIC design.

Check to see that the ASIC vendor includes input protection circuitry on the ASICs to be used. This extra protection will help guard against electrostatic discharge (ESD), which causes the gate oxide to break down, making the device fail.

Design for Testability (DFT)

Testability is the key to achieving manufacturable designs. Design for testability revolves around the principles of observability and controllability for the design. The topics in this section will be covered in more detail in Chapter 8, where details of test considerations and test methodologies are discussed. This section will provide some quick reference notes to assist with achieving a high level of testability for a new ASIC design.

Use combinational logic over sequential logic

Using circuits where the outputs are always a function of the inputs (combinational) allows a design to be more easily tested. Avoid as much as possible circuits where the outputs are dependent upon the current inputs and the current outputs (sequential).

Another recommendation from ASIC vendors is to use synchronous methods rather than asynchronous circuits. Asynchronous circuitry is susceptible to race condition and hazards, which also affect reliability. Race conditions will vary with processing, which means that they will not always appear as part of a design simulation. To ensure that all logic follows synchronous operations, make sure that all feedback paths are clocked to avoid any signal hazards.

DFT and storage devices

Since today's complex ASIC designs usually contain some storage circuitry, some form of initialization is necessary to control the state of the sequential circuit. Most ASIC vendors recommend the use of dedicated test logic to guarantee controllability. Storage elements such as flip-flops, latches, and registers must have an initialization control signal such as a reset to set them to known states for testing.

In addition to initialization, the design may require extra cells to achieve observability. In some configurations, a multiplexer circuit could be used to observe the outputs of a set of flip-flops.

Some ASIC vendors recommend that designers not use their own storage devices. The primary reason is that user-designed storage devices do not have known layout specifications, which will cause timing problems and makes the ASIC less reliable. If the design calls for a specific storage device that the ASIC library does not have, contact the vendor to see about having one designed for the application.

Partition sequential counter chains

Multistage counters should be partitioned into smaller sequences to allow the circuit to be controlled more easily and to reduce the amount of test vectors required. Make sure that each section can be preset separately so that the sections can be tested in parallel, using the same set of test vectors.

Use scan-based techniques to achieve testability goals

Scan techniques are being widely adopted for ASIC testability because of their ability to provide complete observability and controllability for ASIC circuits. The basic operation is to connect internal latches to form a serial shift register, which can be multiplexed to bit-shift data into and out of the ASIC. One of the methods for implementing scan design is level-sensitive scan design (LSSD). The LSSD method uses a serial scan approach to partition the design around latches (levels) and groups of combinational logic where each latch (level) can be controlled separately or as part of a group. This method allows the data flow to be easily controlled through each stage, or level, of the design. LSSD techniques require all latches in a design to be replaced with a data-multiplexed element with multiple control signals. When opting to use LSSD, the designer must follow stringent design rules to avoid any race conditions or hazards. If LSSD rules are followed, LSSD circuits will not be affected by normal AC circuit characteristics. LSSD is not as easy to implement as other scan techniques because of the specific rules that must be followed.

Scan path is another scan technique. It operates on the same principle of shift register sequences but does not rely on partitioning the circuit into levels. Scan path relies on a single clock to control the serial shift chain and can be affected by race conditions. Designers using scan path should use synchronous operations and pay close attention to synchronous design rules to avoid problems with scan path techniques.

These techniques will be discussed further in Chapter 8, "Test."

SILICON COMPILERS AND ASIC DESIGN

Silicon compilers provide a unique capability to compile (or translate) high-level functional (behavioral) design descriptions into lower-level structural functions that are optimized for simulation, timing verification, or layout. Silicon compilers can also perform die-size and power estimates for a specific design. The main benefit of the silicon compiler is that once a design's architecture is selected, the compiler can provide layout details needed to fabricate a complete design. An additional benefit is that the silicon compiler will optimize the design based on specified parameters and design constraints.

The use of silicon compilers has been concentrated in cell-based design for full custom ICs. Cell-based design techniques have become the mainstay of ASIC design. With cell-based design being used for complex ASICs, making ASIC design similar to full custom IC design, silicon compilers are beginning to play a more significant role as the design optimizer for achieving the most efficient layout densities. A perceived drawback of silicon compilers is that they are vendor- and process-specific. The silicon compiler solution will be customized for a specific technology and ASIC vendor. This limits the designer's flexibility and forces him or her to make technology decisions too early in the design process.

Silicon compilers come in different versions that allow a designer to target a specific design to be created. Silicon compilers are divided at the first level into structural and functional (behavioral) compilation types. Each type can be further subdivided into specific unit types that produce results based on the desired design parameters defined by the designer.

Some EDA companies offer silicon compilers that provide a range of design applications: high-level description decomposition, schematic capture, logic simulation, timing analysis, test vector generation, and layout.

The use of logic synthesis techniques for complex ASIC design is increasing. A natural fit for silicon compilers and logic synthesis capabilities is to use a silicon compiler to generate logic modules, then use logic synthesis to derive the structural (gate-level logic) details for those logic modules. This tight integration provides the ASIC designer with more efficient designs and improved productivity and shortens the implementation time.

At ASIC vendor design centers, silicon compilers can be used with custom compiler techniques to provide compiled macrofunctions (such as function generators) for ASIC cells. This allows the designer to concentrate on the constraint parameters for the cell design, not on how to use the silicon compiler.

What Can Silicon Compilers Do?

Silicon compilers can take a high-level behavioral circuit description and compile it into a format ready for simulation or layout. Silicon compilers have the ability to control design parameters for a device to optimize it for density, speed, and/or power.

Silicon compilers provide more efficient cell compaction and aid designers in obtaining the best layout densities in full custom integrated circuit design applications.

Silicon compilers can be used to interject uniformity into VLSI circuit design while allowing a designer to move quickly from an architectural or behavioral description to a complete design.

ERC AND DRC CONSIDERATIONS

Electrical rule and design rule checks are essential for determining whether a design will meet specifications for doing the job desired. ERCs apply to simulations and actual circuit functions. DRCs apply to actual boundaries and guidelines for creating a design. ASIC designers should perform ERC and DRC checks before running any simulation, as these checks will save development time, reduce analysis complexity, and reduce anxiety during analysis of complex ASIC designs.

ERC and DRC checks are sometimes included as part of the EDA vendor's post-schematic capture processing procedure. For the most part, ERCs and DRCs are developed by the ASIC vendor to ensure that new ASIC designs meet their manufacturing requirements. ERC and DRC checks are performed as a postprocessing procedure after schematic capture is complete. ASIC vendors will not allow this process to be skipped. A trend is for schematic capture tools to do vendor-specific ERCs and DRCs interactively.

The following sections will explain the differences and definitions of ERCs and DRCs in more detail.

ASIC AND EDA VENDOR DESIGN RULE CHECKS

ASIC and EDA vendors have developed a vast array of ERC and DRC programs that will ensure that a design can be simulated, laid out, and manufactured using their technology and processes. DRCs that an ASIC vendor needs to verify are parameters related to layout, with a focus on those used to specify structural width and spacing violations.

Designs are checked for conformance with electrical rules and design rules at the end of a schematic capture sequence, prior to simulation. A number of these checks are shown in the list below, which was developed from ASIC and EDA vendors' lists.

Dangling nets
The number of external pins
Instances without pins
Floating input pins
Naming convention violations
Unconnected pins

Vendor-required properties

Bus structure violations

The number of cells

Fanout violations

The number of nets

These checks typically cover the basic rules the EDA vendors require to ensure correct processing between tools. More ERCs are performed after exiting the schematic capture tool. Specific processing tools are used to determine electrical parameter violations such as loading and fanout. In most cases, users can add their own ERCs to the EDA vendor's. This is done by modifying or creating a command file containing the desired checking.

For custom IC design, the EDA vendor will include additional, more stringent checks that ensure layout conformance. Refer to the ASIC vendor's data book or design kit documentation for specific rules that must be followed. Each vendor will have a slightly different set of rules to fit its own processing and technology offerings.

KNOWING WHEN THE DESIGN IS READY FOR VERIFICATION

How does a new ASIC designer know when the design is complete and ready for the next process step? The ASIC vendor will usually provide a set of guidelines to let a designer know when the design is ready to be sent to the foundry for prototype manufacture.

All functionality entered

Design specifications complete

Library sources identified

Schematic symbols completed

Design compiled for layout

Netlist created for the ASIC vendor

Logic and timing verification complete

Critical timing circuits identified

Test vectors developed and graded

Package type, bonding diagram, and pin ordering determined

All forms completed for ASIC vendor processing of the design

System design team review completed and signed off

ASIC vendor design review completed and signed off

SUMMARY AND CONCLUSIONS

Design creation is the most important step in making a high-quality ASIC that works the first time. Designers must take the time to define detailed specifications, observe (via simulation) high-level operations to verify that the concept will work, and choose the right tools for the job relative to time to market, costs, and expertise.

Industry research has shown that a top-down approach to complex ASIC design is the most effective development method. Design creation is more than just performing schematic capture. All of the design's constraints, dependencies, and functionality must be clearly understood before embarking on schematic capture.

Choosing the right level at which to begin the design process can make the difference between hitting or missing the market window for a new product. Industry researchers have shown that over 80% of the cost of a new design is determined in the first 5% of the design process. Furthermore, as the product advances in the design cycle, the costs for making changes increase exponentially to over a factor of 3.

There are many versions of schematic capture packages that can be used to design an ASIC. Designers must choose between buying their own tools, using an ASIC vendor's, or using a distributor for developing the new ASIC design.

There are several tools that can help smooth the development process for designers with their first ASIC. VHDL, logic synthesis, and silicon compilers offer a designer ways to develop accurate, high-quality ASIC designs at the lowest cost.

REFERENCES

CE/MG Co-designer Environment for the Mentor Graphics System. LSI Logic Corporation, 1989.

Gajski, D., and R. Kuhn. "Guest Editor's Introduction: New VLSI Tools." *IEEE Computer,* December 1983.

Kane, R., Sahni, S. A Systolic Design-Rule Checker. *Computer-aided Design of Integrated Circuits and Systems,* Volume CAD-6, No. 1, 1987.

Logic Synthesis Can Help in Exploring Design Choices. *Semicustom Design Guide,* 1989.

Mentor Graphics Workstation Users Manual. VLSI Technology Inc., 1988.

NCR ASIC Databook. NCR Corporation, 1989.

Thomas, D. E., Blackburn, R. L., and Rajan, J. V. Linking the Behavioral and Structural Domains of Representation for Digital System Design. *Computer-aided Design of Integrated Circuits and Systems,* Volume CAD-6, No. 1, 1987.

4

Design Methodologies for ASICs

INTRODUCTION

ASIC designs can be developed using several different methods. The method chosen depends on a combination of the designer's experience, the design's size and complexity, the technology needed for the design, and the processes the selected ASIC vendor (or foundry) supports (see Figure 4-1). To a lesser extent, the method chosen can depend on whether the designer has EDA equipment in-house or the design must be performed at the vendor's design center.

In today's complex ASIC design environment, a designer's choices for ASIC development are limited to two ways:

- Bottom-up: Starting at the structural level and adding hierarchy to the design based on structural components and limited hierarchical models consisting of behavioral constructs.
- Top-down: Starting with a functional representation of the design, then detailing each section of the design by partitioning until a structural representation is realized.

This chapter will discuss the details of these methods as they apply to ASIC development. For each method, there will be recommendations and

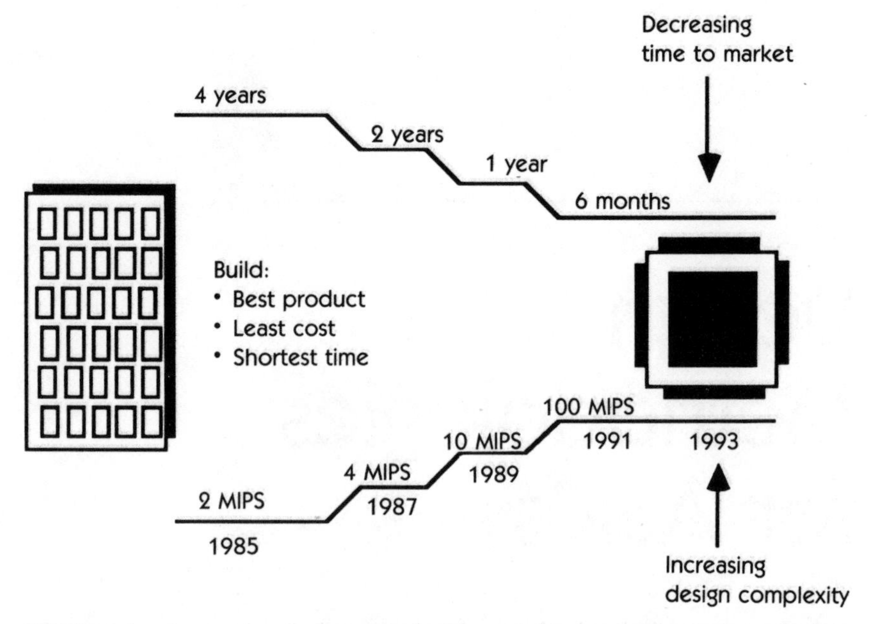

FIGURE 4-1. Increasing design complexities are forcing designers to seek alternative design methods. *Courtesy of Mentor Graphics Corporation, 1990.*

suggestions about when each approach should be considered. For example, there may be times when a silicon compiler approach should be considered as part of the top-down development process to assist the designer in optimizing the ASIC for speed, area, and power consumption.

BOTTOM-UP DESIGN

Bottom-up design has been the traditional method for design of ICs and printed circuit boards, once the designer proceeded past the concept stage. Bottom-up basically means to begin building the circuits at the structural level by selecting and connecting cells or logic gates. This bottom-up method was fairly effective until ASIC designs began to approach a complexity level of 10,000 gates. Designers then were not able to cope easily with the structural-level details. If the design method did not change, designers would be faced with longer product completion times, lower reliability, and higher development costs.

Complexity was masked somewhat by adapting a hierarchical design approach to structural-level design (see Figure 4-2). This method consisted of building a symbol over a structural block in a design. Then the

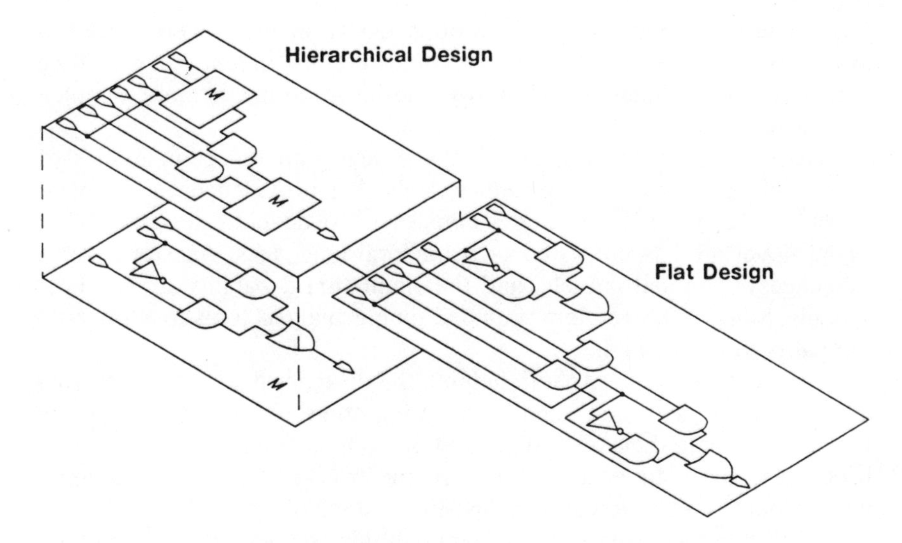

FIGURE 4-2. Hierarchical schematic example.

blocks could be combined into larger groups using a new symbol to cover the detailed functions. Designs ranged from those having as few as two levels to those with more than six layers of hierarchy. The advantage of this approach was that the designer was able to deal with most of the design using functional-block representations and drop down to the structural level only during simulations when performing debugging of the design.

A typical design cycle consists of:

1. Perform a conceptual pass on paper.
2. Manually partition the design into major functional blocks.
3. Perform functional simulation to verify specifications.
4. Build structural schematics of each functional block and reverify.
5. Define pinouts and bonding diagrams.
6. Verify each of the blocks using logic simulation.
7. Verify the entire design with logic simulation and timing analysis.
8. Develop a set of test vectors.
9. Send the design netlist to the ASIC foundry for layout.
10. Test the prototypes.

Because of complexity issues, other design methods were forced to appear. EDA vendors and ASIC vendors were instrumental in providing new tools and techniques that allowed designers to tackle more complex chip designs.

EDA's role in helping the designer deal with the complexities of structural designs was to introduce tools that provided designers with ways to use higher-level model constructs to develop designs. These higher-level models consisted of behavioral languages, register-transfer languages, compiled models, and the simulation capability to use these models. Some EDA vendors provided timing analysis tools to allow critical paths to be verified.

The ASIC vendor's role in helping the designer deal with increasing design complexities consisted of adapting some of the EDA vendor's tools, developing their own tools, and bringing new technologies to their processes. Early to be adapted were the front-end tools, technology-independent schematic capture, and logic simulation tools.

ASIC vendors also began to adapt their libraries to the EDA vendors' front-end development tools. This allowed ASIC designers to use their own workstations and to use the ASIC vendor's library, which was qualified for the EDA design tools. ASIC vendors could then sell their libraries to designers and manufacture the completed ASICs.

TOP-DOWN DESIGN

Of the two methods, the top-down method is preferred, especially when ASIC design complexities exceed 10,000 gates. Top-down design begins with the high-level design specifications and requirements. The designer then partitions the design into functional blocks with detailed timing specifications for I/O. The functional blocks are then turned into behavioral or register-transfer models, depending on their function, and simulated. After functional verification proves satisfactory, a block is decomposed into structural logic, simulated, and the timing verified. The entire design is decomposed into structural-level logic blocks and verified against the system specifications.

The advantages to a top-down design approach are basically:

- It is a natural design method to use for meeting specifications.
- Designers can execute design operations at the system architectural level using behavioral constructs without being encumbered with structural details.
- Designs can be validated independent of the technology to be selected, allowing the technology choice to be made after the final structural schematic representation is performed.

- Design evaluation can actually take place at each step of the design's evolution from behavioral to structural representation.

This chapter explores ASIC design using a top-down methodology. The rest of the book assumes a hierarchical, top-down approach to the design of complex ASICs with over 10,000 gates. The top-down method is the recommended approach for complex ASIC design.

CONCEPTUAL-TO-FUNCTIONAL DESIGN

As we discussed in Chapter 2, the place to begin a complex ASIC design is with conceptualization and detailed specifications. This stage may extend to defining high-level functional blocks that show datapaths and I/O. At this juncture, the designer has several options:

1. Take one functional block at a time, develop structural logic equivalents, and verify the design against the specs.
2. Take several functional blocks, turn them into structural logic equivalents, and verify the design against the specs.
3. Replace all functional blocks with structural logic equivalents and verify the design against the specs.

Whichever way the design effort is approached, after partitioning the process will be iterative. The important point is that the designer should never move to the next phase of the process until he or she has verified that the design meets specifications.

The most efficient method for top-down design of ASICs is to begin with an abstract description of the circuits. There are several standard methods used for abstract descriptions: behavioral language modeling, register-transfer language modeling, and the newest comer, VHDL, described below.

Complex Cells Offer a Step in the Right Direction

Most ASIC vendors offer several levels of predesigned cells that a designer can use. The level of complexity and the number of complex cells offered will vary from vendor to vendor. The designer can choose these higher-level macro functions and simply connect them together. Cells in this category contain counters, ALUs, multiplexers, and core microprocessors.

This level of cell complexity offers the designer some relief from structural-level details for some types of designs. However, it does not free the designer from the checks for accuracy and performance.

VHDL—IEEE-1076 Standards

Thanks to the Department of Defense (DoD) and the very-high-speed integrated circuits (VHSIC) project, a standard for hardware description languages (HDL) has been established. The standard, IEEE 1076-1987, has been widely accepted as the best approach to complex systems design. The use of VHDL allows designers to describe whole ASICs at an architectural level, which is a high-level abstraction of a structural circuit.

The DoD has mandated that all ASIC designs submitted under contract must be accompanied by VHDL models based on IEEE-1076 as a part of Military Standard 454. Military Standard 454 further declares that "ASIC designed after September 30, 1988 shall be documented by means of structural and behavioral VHDL descriptions."

The fundamental reason for this push to VHDL is that as designs get larger, the structural representation of logic gates becomes too complex for designers to contend with. Therefore, describing a design at a higher level of abstraction is expected to provide a faster means to achieve optimum power consumption, performance, size, and costs. Further, using VHDL will provide higher-quality ASICs while identifying design flaws earlier in the design cycle. The goals behind the standard are:

- To provide a standard way of documenting design descriptions.
- To provide a standard high-level, abstract simulation input language to describe models that can run on any logic simulator that supports VHDL.

There are two camps with regard to this latter goal. One is those who believe that if they support only a subset of VHDL, their models support VHDL, the "standard." The other camp believes that they must support the full VHDL standard; otherwise they cannot truly claim VHDL compatibility. The underlying assumption here is that VHDL methods for high-level models should allow those models to be interchangeable among simulators that support VHDL. An example would be a designer who designs models for an ASIC that gets sent to an ASIC vendor. The ASIC design should be able to be simulated at the VHDL level if both simulators support the VHDL standard. This is something that many believe will never come to pass because simulators are not standardized. In Chapter 5, the story of digital simulators will be told, and reasons why there may be a need for a simulation standard to match VHDL can be seen.

A major problem in the industry is that there are no standards on how to validate full or partial VHDL support. Several vendors provide various "VHDL-compatible" models, but there is no standard way to test them. Normally, simulation models are defined for a specific simulator. If a vendor is said to have a VHDL simulator, will someone else's VHDL models work? That is the goal, but since there is no standard simulator, it is not clear that it will be possible for VHDL models to be interchanged between simulators.

What is still needed, therefore, is a standard test suite to provide full-scale validation of a vendor's compliance with IEEE-1076. Several organizations are working to provide a standard test suite with which to measure VHDL compliance.

The best advantage of VHDL for ASIC designers is that it provides an ideal environment for developing a fully functional architectural model of the ASIC's circuitry. This fully functional architectural representation can be simulated to evaluate a design's conformance to specifications and to make a projection of performance characteristics.

A variation on the standard VHDL is the use of a register transfer language (RTL) as a subset of VHDL for describing data flow. RTL, as a subset of VHDL, offers a very efficient way of describing I/O interfaces, with a VHDL architectural declaration then used to define the function. The RTL variation allows the designer to more easily define data flow, especially transactions taking place on bus structures.

Now, since an abstract representation of a design does not have any structural definition, there can only be limited timing verification at this level. This means that sooner or later the designer has to turn the abstract description into structural logic in order to completely verify how reliable and accurate the design is. The only way to complete accuracy measurements for an ASIC design is to use a structural-level view to verify input and output delays for each gate in the circuit. These delays can be used by a timing analysis tool to determine critical paths, setup violations, and clocking violations. There will be more on timing analysis in Chapter 6.

The VHDL representation can be changed into structural logic through manual design or using design synthesis techniques. By using design synthesis techniques with VHDL, designers can quickly decompose VHDL descriptions into structural logic and perform detailed verification of complete ASIC designs.

DESIGN SYNTHESIS

Complete automation of the ASIC design process would allow a designer to conceptualize a design, define the constraints, push the "green button," and then wait for the prototype ASIC to appear. Sounds far-fetched, right? Synthesis proponents have been predicting this scenario for the last several years. Interestingly enough, design synthesis does offer a solid basis, albeit only a glimpse, for realizing the automatic design of complex ASICs.

Basically, design synthesis is defined as having two levels or distinct processes, logic synthesis and layout synthesis. Sometimes there are references to a third process, behavioral synthesis. For our purposes, behavioral synthesis is included as part of the logic synthesis process because of commercial product offerings. These distinct processes are used to set up the design for iteration and optimization during the course of each step. The amount of iteration and optimization needed is solely dependent on how well the design meets its specifications. We shall explore the details of both methods of design synthesis in this section and discuss how each applies to ASIC design.

Design synthesis brings with it questions that a designer and a design manager must ask to decide whether it should be used on their new ASIC project (see Figure 4-3). Some of these questions are:

- Will the chip function as specified with other chips in the target board? How can this be guaranteed?
- How will the designer know whether a defined high-level function will be guaranteed to meet specifications for a particular technology?
- Will the designer be able to make trade-offs between different technologies (at the abstract level) to solve predicted timing problems?

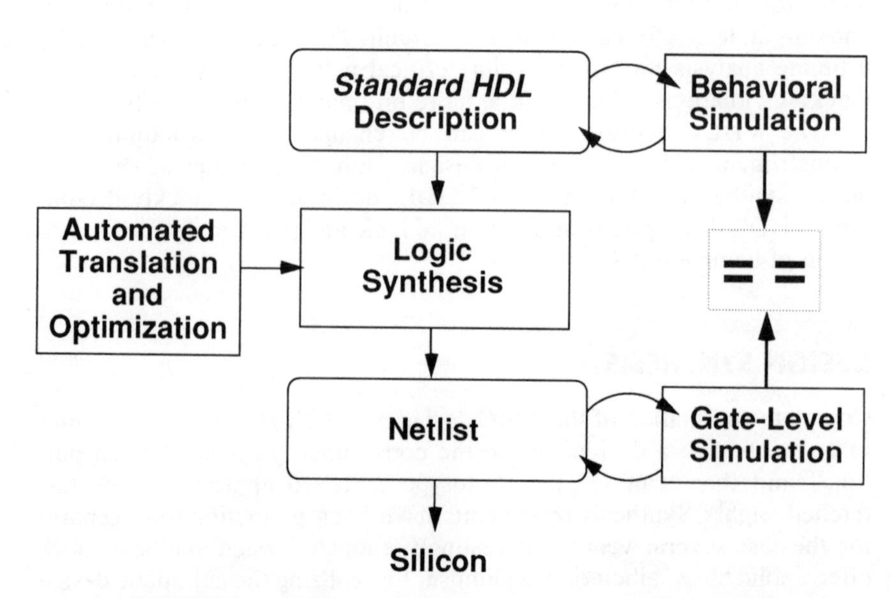

FIGURE 4-3. A logic synthesis cycle description. *Courtesy of Synopsys Incorporated.*

- Can the designer make accurate predictions of chip size and production costs and trade off at an abstract level?

Some of these questions can be answered only by going through a complete design synthesis process and measuring results at the structural level. Further, the answers may require several iterations with the design synthesis process to get the design correct. Other answers lie in performing full simulation at an RTL or behavioral level.

The ability to quickly synthesize the design to the structural level allows design trade-offs for performance or technologies to be more easily made. Using these techniques, designers can make rapid checks to determine chip size and cost analysis for different technology approaches.

Does the entire design need to be synthesized, or can the designer just operate on portions of the design? Using silicon compiler techniques can save time and trouble, especially in RAM-based designs, because of the repeatability of each cell. Normally a design is partitioned into functional blocks for ease of verification. Using design synthesis to develop structural representations of a single block and then simulating as a full system can often provide a faster means of verification.

With design synthesis techniques, the designer is often faced with a decision about which level should be used to optimize the design. Should designers optimize the structural-level descriptions or optimize at the VHDL or behavioral level? The right level for circuit optimization depends on several things: the tools being used, the experience of the designer, and the function of the circuit.

It is important to know that some timing analysis tools cannot provide accurate timing checks of components described in higher levels of abstraction. Also, some synthesis tools cannot offer accurate synthesis from VHDL models to provide exact structural logic representations. Some synthesis tools do not provide knowledge bases for developing design or process methods that can guide novice designers.

Silicon compiler tools are quite complex and are not designed with the novice designer in mind. One class of silicon compilers is targeted at the custom integrated circuit designer who understands the intricacies of IC layout. The other class of silicon compilers is targeted at the system designer who does not have a lot of experience with IC layout and may not want to perform design verifications and DRCs. The important point for the new ASIC designer who may want to try silicon compiler technology is to know which category he or she is in and which class of silicon compiler is being used by the chosen ASIC vendor.

Synthesis techniques cannot currently provide a way to handle mixed analog-digital behavioral constructs and convert them to a structured form. Mixed analog-digital synthesis is certainly an area that must be targeted for future development efforts.

LOGIC SYNTHESIS

Logic synthesis techniques try to guarantee that designs that are developed at the architectural level will meet their requirements in the structural logic version. Simply put, logic synthesis is a process that automatically generates the structural logic equivalent of a behavioral circuit description. These circuit descriptions are used for digital simulation to verify how well the design meets specifications. The design may be simulated with different sections of the ASIC defined at a mix of the levels mentioned above. There may be sections at the architectural level mixed with an RTL segment that is tied to a full structural representation.

Design synthesis techniques normally include a knowledge base that can interact with the design netlist and constraints lists to attempt to optimize the design. Design synthesis techniques that use a knowledge base are known as expert system–based synthesis tools. A knowledge base consists of a set of rules and techniques (algorithmic) for using the design data structures and making modifications to the data structures.

Design synthesis techniques built using expert systems fall into one of three categories: rule-based, rule/algorithm-based, or algorithm-based.

Rule-based systems generally lack order or structure, as rules are entered independently of one another. Generally, a rules-only synthesis system will require a specific conflict resolution scheme to ensure that the right rule is used.

A combined rule/algorithm approach provides a control mechanism over the rules that must be considered at a given time. This type of synthesis system can introduce a hierarchy to reduce the number of rules that must be considered at a given time. Techniques are employed to limit rule search time by prespecifying a certain rules order that the synthesis optimizer must follow.

The algorithm-only synthesis system means that the optimizer will always perform the same operations in the same order.

A knowledge-based synthesis tool will guide a new ASIC designer through the top-down design process. The guidance is provided by relying on the rules contained in the expert system knowledge base that tell the designer and the optimizers which decisions to make for the best design.

When knowledge bases are used to their fullest extent, novice designers benefit from the experience of senior designers. This benefit is provided through the ability to use the senior designer's trade-offs as some of the rules embedded in the knowledge base. Based on their many years of experience, ASIC vendors who employ knowledge bases should be able to provide the best leverage to help new designers with the successful development of new ASICs.

It is interesting to note that all knowledge-based systems using a rule-based approach to design can only reflect the abilities of the individuals who were involved in establishing the rules for the system.

Whether the design is done in-house or at an ASIC vendor design center, today's device complexity means that new ASIC designers should take advantage of all the help that is available. This complexity affords little room for mistakes.

SILICON COMPILERS

Silicon compilers are a relatively new technique that came out of university laboratories and into the industrial segment in the early 1980s. Much of the credit for getting silicon compilers into industry goes to Carver Mead, who is known as the father of the silicon compiler.

The first silicon compiler to be offered commercially was by VLSI Technology in 1983. The VLSI silicon compiler was *technology-independent* but ASIC vendor–dependent. Of course, VLSI was the vendor whose process the silicon compiler tracked. The first *foundry-independent* silicon compiler was introduced in 1984 by Silicon Compiler Systems, now a part of Mentor Graphics Corporation. Those early silicon compilers were limited in their capabilities and were used only to generate RAM, ROM, and PLA blocks.

The silicon compiler's greatest contribution to the industry is the reduction in chip size. This reduced chip size is due to the silicon compiler's ability to use silicon efficiently. The reason that silicon compilers were so popular for building memory structures was that since all transistors are uniform, the packing density could easily be maximized. Chip utilization by a silicon compiler is better than that of either gate arrays or standard cells.

Silicon compilers are used by a limited number of ASIC vendors. Some of the current ASIC vendors using them are LSI Logic, NCR, Texas Instruments, Motorola, VLSI, and Plessey. Interestingly enough, they all use different silicon compilers.

The key to silicon compilers is their ability to provide module generators, custom and foundry. Module generators are the specific compiler output used to produce a specific logical function. Module generators provide functions such as counters, multipliers, RAM, ROM, and ALUs. These module outputs are developed according to vendor specifications and placed in the foundry's library for use by ASIC designers.

Silicon compilers perform optimization of functional, performance, and physical characteristics for ASICs; they allow optimizations for speed, power, and area to be controlled through user-defined constraints;

and they also allow designers to obtain accurate feedback about how well the design is meeting performance goals.

Types of Silicon Compilers

There are several types of silicon compilers: random logic, datapath, module compilers, and tile-based. While each is designed to provide a chip layout mask, each type offers unique characteristics for specific design rules. A general benefit of silicon compilers is to offer the user various means for defining design rule parameters. These design rule parameter inputs can be algorithmic, menu-driven, schematic-driven, or netlist-driven.

One of the earliest silicon compilers used an algorithmic input method. Use of algorithmic inputs did not track well with EDA design methodologies and has not been as popular as the use of menus, netlists, or schematics.

A random logic compiler is used to develop gate arrays and standard cell designs that consist of logic gates, flip-flops, buffers, and latches. Datapath silicon compilers are used for developing multibit data configurations in a bus architecture. The datapath silicon compiler is useful in applications for graphics controllers, digital signal processors, computers, and computer peripherals. Module compilers are available for special, even custom blocks such as RAM, ROM, and PLA. Tile-based compilers perform compaction layout of blocks arranged as "tiles."

Silicon compilers must be calibrated to achieve optimum results for a particular set of design rules. Since silicon compilers are closely tied to the vendor's process, calibration will be performed by the ASIC vendor (or the ASIC process engineer in an internal ASIC group) using their process parameters. Further, designs created using a silicon compiler should be verified using layout versus schematic comparisons. Experts recommend that the designer understand calibration of a silicon compiler if the ASIC vendor of choice used one for developing process cell models. Knowledge of the process will enable the designer to gain a better insight into any custom cells required and to be satisfied that the accuracy of the cells being used is satisfactory. The designer should verify whether the silicon compiler was calibrated and, if so, for which process technologies.

Checking on the silicon compiler used to determine whether custom modules are easy to generate is highly recommended. The ease of module generation is determined in part by whether the silicon compiler includes options that allow the ASIC vendor to perform the necessary calibration steps to certify the accuracy of a cell.

The basic calibration process (see Figure 4-4) consists of entering the design rules and process parameters, running DRCs, checking for errors

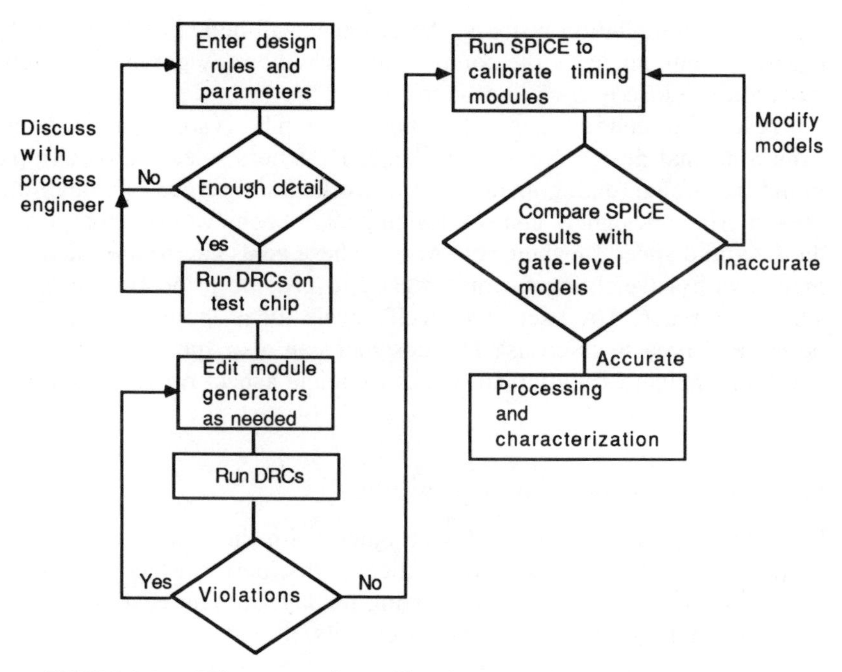

FIGURE 4-4. Silicon compiler calibration steps for module generation.

and running SPICE simulations to calibrate timing modules, comparing the results with gate-level models for logic simulation, then processing and physical characterization of test wafers. If all checks out, a new module generator is developed for the ASIC vendor, resulting in a new cell for use by ASIC designers.

Let us take a look at the details of using a silicon compiler and discuss how it applies to optimizing functional operation, performance, and physical characteristics for an ASIC design. The important thing to remember with silicon compilers is that they are silicon process–dependent and that the function cells generated are targeted for a specific ASIC vendor.

Silicon compilers are accessed through schematic capture or by using pop-up forms as part of the design creation process. The details of design creation were explored in Chapter 3.

For a specific module to be generated, the designer must input all constraints that are important to the overall design. Default settings can be selected for other parameters. The whole idea is to provide a completely optimized design based on user-defined constraints such as speed, power, and area. A silicon compiler will automatically seek to optimize the design in such a way as to best fit the specified constraints. At each step in the compilation process, the designer will be kept advised of the progress being made by the compiler and whether any parameters are in violation or close to violation boundaries.

As can be seen in Figure 4-5, a silicon compiler is somewhat different from a normal design creation package. Designers select or specify parameters through dialog menus on the workstation screen. Through this user interface designers can set design goals to achieve optimum power, die size, and speed for their new ASIC. These goals can be specified as a matrix with a weighting factor assigned to provide prioritization to the silicon compiler. The user can specify cycle times, power in microamperes, and area in microns. The designer can also specify the relative width and height of individual blocks to define aspect ratios for layout.

Functional Optimization Techniques

Functional optimization provides designers with the ability to define an ASIC cell structure to determine critical paths, perform logic minimization, and provide detailed cell mapping for layout. The basic goal is to minimize gates, removing redundant circuits.

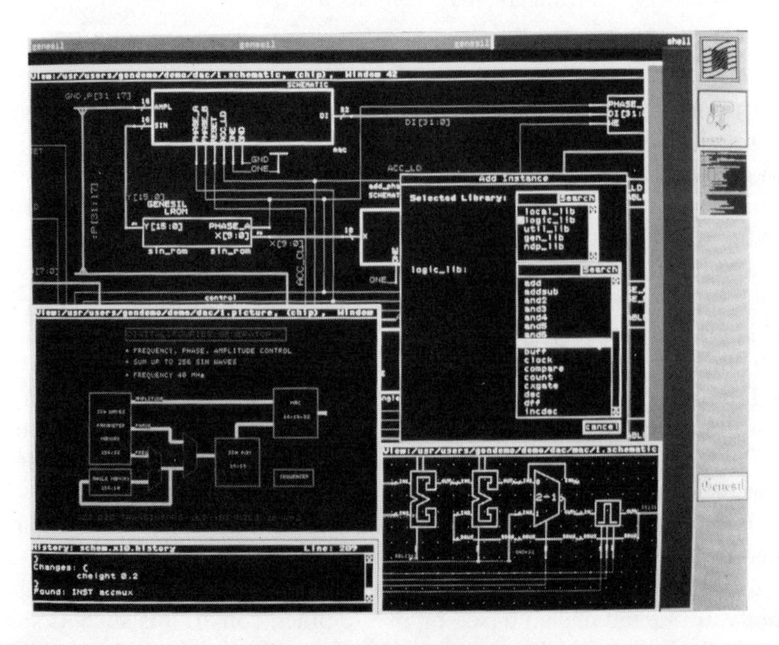

FIGURE 4-5. Photo of a screen shot for GENESIL silicon compiler from Mentor Graphics.

The first stage results in a functional simulation model based on simulation primitives like AND, OR, XOR, NOR, and latch. The next step transforms the functional simulation model into a process-specific cell.

For cell mapping, the compiler will perform two steps to ensure that an accurate model is developed. The first step will be to take the source netlist from the schematic entry and develop a simulation model. The next step will translate that functional model representation into a process-specific cell.

The transfer of the logic function into a cell primitive can be different, and the resulting cell is determined by the target process. This means that for a NAND gate function, the characteristics of performance, area, and power consumption will differ from those required for a NOR gate, for example.

Logic minimization and optimization are accomplished through a technique comparing the newly defined cell parameters and any similar definition resident in the ASIC vendor's database. Functional optimization can include the use of artificial intelligence to learn about the fundamental design rules of a cell being used in a designer's circuit. If the new cell has better specifications for the application, then it is used and can even be added to the existing database. If, on the other hand, an existing cell provides the same or better specifications than the new cell definition, the existing cell is used instead. The new cell will be discarded.

Critical path optimization can be automated by the silicon compiler as well. A silicon compiler will attempt to optimize the delay in critical paths in order to provide the smallest delay that will meet design constraints. If extra logic is created, the logic optimizer will automatically perform logic minimization on the new circuits. These automatic techniques allow a designer to use a less than optimum logic design and let the compiler perform all circuit refinements. The danger with this is that the design may change so much that during design verification and debugging, the designer may not recognize the original design.

Performance Optimization Techniques

The function of performance optimization techniques is to select the proper size transistor for each gate in the circuit. Depending on the user's requirements, a silicon compiler can use performance optimization to tune the circuit for speed, area, and power trade-offs. Trade-offs for speed can be achieved by making certain transistors in a critical path larger. However, the power and area will also increase. The silicon compiler must be intelligent enough to determine which transistors can be increased and which can be minimized to optimize all critical paths. This optimization is accomplished by having a high-speed estimator to ensure that the design compilation will converge to a solution in a reasonable amount of time. The time it takes is a function of the workstation platform used and the size of the design.

While this step does optimize for timing, the designer should not consider this a substitute for static timing analysis of the ASIC design. A thorough discussion of timing analysis will be left for Chapter 6.

Layout Synthesis—Physical Optimization Techniques

Silicon compilers optimize the physical layout of the ASIC using as inputs the parameters derived from the functional and performance steps. The sizings for transistors are input to produce optimum placement and routing. Using synthesis techniques, area usage reductions can be as much as 20 to 30% depending on the algorithmic techniques used.

These techniques include the ability to compute timing-driven placement of the cells. The inputs from a static timing estimator are used to calculate cell locations to minimize interconnect routing and still maximize performance. This technique allows an optimizer to dynamically size all transistors along a critical path to meet timing constraints. At the same time, transistors that are off the critical path are minimized in size to keep the design within the defined area constraints.

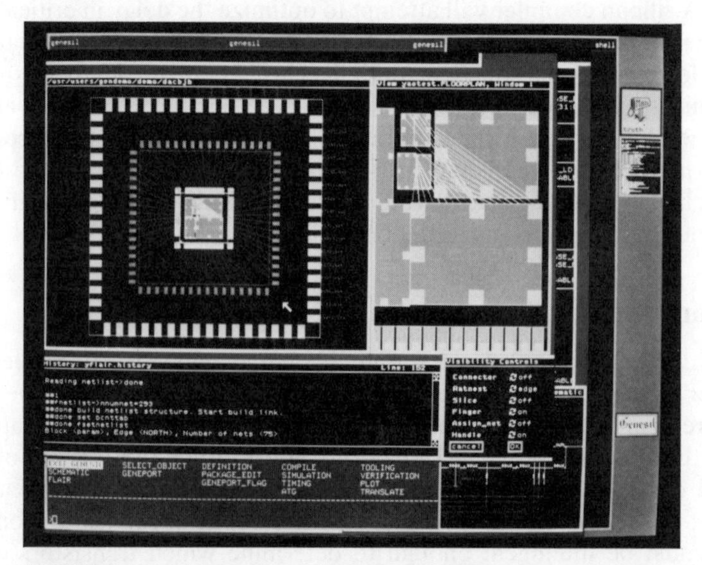

FIGURE 4-6. Placement and 100% automatic routing are an integral part of GENESIL Designer's floorplanning system.

A different algorithm can be used to compute ways to minimize area by clustering cells that have a high interconnectivity requirement. Based on the algorithms used, the silicon compiler may allow the user to choose to move clusters of transistors, clusters of cells, or single cells to optimize the design.

A final step for layout optimization with silicon compilers is to determine the sizes for all transistors for each gate in the design. This is referred to as electrical optimization. A silicon compiler can use the weighting factors assigned by the user to track the performance lost versus area saved for every transistor size change in the design. The silicon compiler then can make intelligent trade-offs for design optimization by weighing the area savings of smaller transistors against the performance gained by using larger transistors for each gate in the circuit. This trade-off analysis is especially useful with critical paths in the design by allowing those transistors that provide the greatest speed improvements along with the least costly critical path to be changed.

With the silicon compiler, the finished process is geared for layout according to the rules defined by the designer. These rules are defined in accordance with the design rules and process requirements of the foundry selected by the user. This makes the silicon compiler technique process-dependent. However, silicon compiler technology is rapidly moving toward providing the density of hand-crafted custom integrated circuits.

Silicon Compilers Cannot Do It All

Design synthesis techniques using silicon compilers can be very effective in organizations that use a single ASIC vendor or that have a captive ASIC foundry in the company. This is because the techniques used are technology- and process-dependent.

Synthesis tools that do not require a silicon compiler offer technology independence as well as process independence. This type of synthesis technique allows the designer to postpone choosing the technology and ASIC vendor until later in the design cycle. This technique offers design flexibility and built-in vendor independence, providing a quick means for second-sourcing an ASIC.

Like silicon compilers, other design synthesis tools provide the means to synthesize a design by functional blocks. Design synthesis also provides the means to reuse those functional blocks for different segments of the design where only parameter changes may need to be made. Reusable blocks save time during the development and verification cycle. Design synthesis can provide a density increase of up to three times over basic standard cells from an ASIC library.

ASIC vendor independence means designing with a generic library and committing the design for fabrication only when it is ready. This also allows the designer to use multiple sources for ASIC designs and not be locked in to a specific process or vendor. This provides an added assurance against any economic uncertainty.

SHOULD THE DESIGNER LAY IT OUT?

Probably not, but when looking at design synthesis for layout, the ASIC designer should be aware of some key issues, especially if he or she will perform some portion of the layout. Knowing about the layout features of the synthesis tool will give a designer some insight on trade-offs to make during the creation and specification process.

Synthesis tools give the designer a lot of control over layout. If an ASIC vendor for manufacturing has already been picked, it is always a good idea for the designer to know the process engineer with whom he or she will be dealing later. The designer should take some time to discuss what he or she intends to do and solicit some recommendations from the process engineer on how to proceed. The tools will ask for parameter inputs for layout. A partial list is:

Block interconnection for identifying chip arrangements

A netlist or set of logic equations

Spacing criteria for transistors

Cell I/O port locations

Clock bus parameters

Device sizes

Information about power bus structure

Process definition for layers

Identification of manual placement structures

Identification of feedthrough locations

For high-density ASICs, the designer must perform an additional step: floorplanning. Because of the complexities of cells, today's ASICs must be planned and even simulated using estimated metallization delays to ensure that the device will work.

SUMMARY AND CONCLUSIONS

In this chapter we explored the various design methodologies for ASIC development that exist today. Different companies may employ variations on the methods described here, but the fundamental procedures are the same.

Designing ASICs with a complexity of over 10,000 gates requires a top-down approach to circuit development. In addition, complex ASICs that include microprocessor cores and support functions should be designed using a top-down development procedure. Further, the implementation should consist of using VHDL-level descriptions of the circuit to round out conceptual design and to solidify design specifications.

Silicon compiler technology has brought a higher level of design automation to ASIC development. Even with silicon compilers, the designer must still have a considerable amount of expertise to be proficient in their use. However, knowledge-based design rules are bringing ASIC design synthesis into the realm of less experienced engineers.

REFERENCES

Baker, A. "Selecting a Silicon Compiler." *VLSI Systems Design,* May 1986.

Barton, D., A First Course in VHDL. Intermetrics Inc., *Design Automation Guide,* 1988.

Burich, Misha R., The Role of Logic Synthesis in Silicon Compilation. Silicon Compiler Systems Corp, *Semicustom Design Guide,* 1988.

D'Abreu, M. Putting Synthesis to Work. GE Corporate Research and Development, *IEEE Design and Test of Computers,* April 1989.

Goering, R. "Design Automation." *High Performance Systems,* December 1989.

Goering, R. "Intelligent Silicon Compiler Optimizes ASIC Design." *Computer Design,* April 15, 1987.

Harding, B. HDLs: A High-powered Way to Look at Complex Designs. *Computer Design,* March 1990.

Harding, B. "Logic Synthesis Forces Rethinking of Design Methods." *Computer Design,* December 1989.

Johannsen, D. L., K. McElvain, and S. K. Tsubota. "Intelligent Compilation." *VLSI Systems Design,* April 1987.

McFarland, M. C., A. C. Parker, and R. Camposano. "Tutorial on High-Level Synthesis." In *Proceedings of 25th ACM/IEEE Design Automation Conference,* 1988.

McLeod, J. What's the Next Step for Silicon Compiler Systems? *Electronics,* July 1989.

Meyer, E. Analog Synthesis: How Feasible, How Soon? *Computer Design,* May 1990.

Meyer, E. VHDL Opens the Road to Top-down Design. *Computer Design,* February 1989.

Miller, T. Does Logic Synthesis Work for Multi-chip, High-gate-count Systems? Integrated CMOS Systems, Inc., *VLSI Systems Design,* October 1988.

Pollack, S., B. Erickson, and S. Mazor. "Silicon Compilers Ease Complex VLSI Design." *Computer Design,* September 1986.

Prabhu, A. M. LOGOPT—A Multi-level Logic Synthesis and Optimization System, AT&T Bell Laboratories, *IEEE Custom Integrated Circuits Conference,* 1989.

Simner, R. "Evaluating Silicon Compilers: Process Calibration, Density, Design Verification." *VLSI Systems Design,* March 1988.

Sullivan, R., and L. Asher. "VHDL for ASIC Design and Verification." *Semicustom Design Guide, VLSI Systems Design,* 1988.

Temme, Karl-Heinz, CHARM: A Synthesis Tool for High-level Chip-architecture Planning, University of Dortmund, *IEEE Custom Integrated Circuits Conference,* 1989.

"Tutorial on High-level Synthesis", Michael C. McFarland, SJ, Boston College, Alice C. Parker, University of Southern California, and Raul Camposano, IBM T. J. Watson Research Center; *25th ACM/IEEE Design Automation Conference,* Paper 23.1, 1988.

Vander Zanden, N., and D. Gajski. "MILO: A Microarchitecture and Logic Optimizer." In *Proceedings of 25th ACM/IEEE Design Automation Conference,* 1988.

Wildman, P., Getting a Grip on Logic Synthesis Tools. Synopsys, Inc., *Electronic Products,* November 1989.

5

Logic Simulation

WHAT IS SIMULATION?

The main purpose of logic simulation is to aid in verification of the correctness of an ASIC design. By definition, simulation is the process of exercising a theoretical model of the design, as a function of time, for some applied input sequence (Breuer and Friedman 1976). The theoretical model consists of a description of how components in the design are connected together, called the *connectivity,* and simulation models that describe the functionality of each of the components. The input sequence is usually referred to as *stimulus* or *vectors*. Connectivity is obtained as a result of schematic capture or a synthesis process; models are normally supplied by the ASIC vendor, and the designer is responsible for developing stimulus that properly exercises the design.

In a practical sense, logic simulation is used to verify that the design meets its intended specification. The designer's ability to use simulation to its fullest will largely determine the confidence he or she will ultimately have in the design and is a factor in moving the ASIC onto the production floor with a high level of quality. Figure 5-1 shows a typical logic simulator.

In addition to logic simulation, fault simulation provides a measure of the effectiveness of the stimulus in fully exercising the design. Fault simulation will be discussed in Chapter 7. Min/max simulation, which provides the ability to analyze a design under worse-case timing, will be discussed in Chapter 6, "Timing."

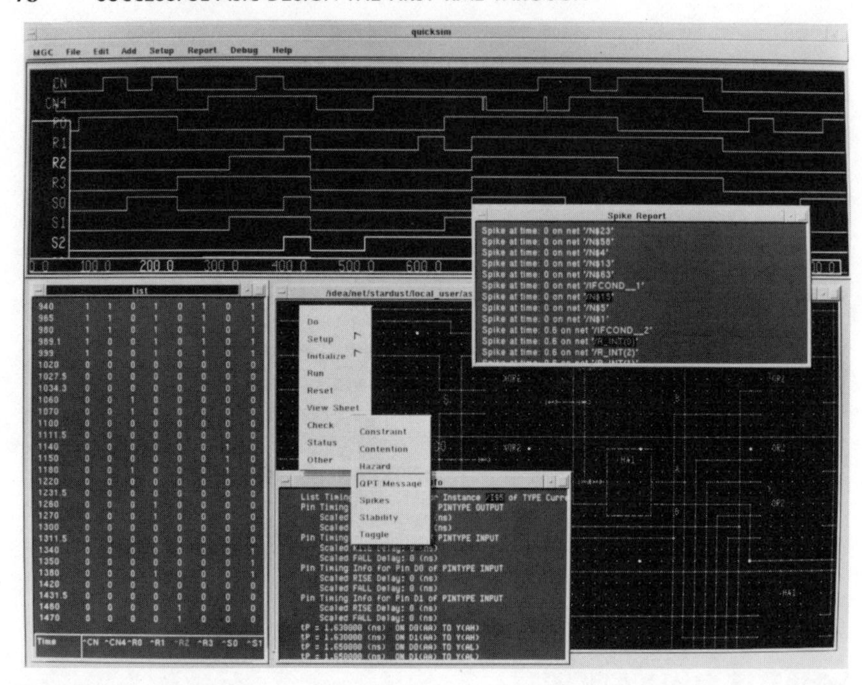

FIGURE 5-1. Quick Sim II logic simulator. *Courtesy Mentor Graphics Corporation, Wilsonville, Oregon.*

Logic simulation results are almost always required as part of the package of information submitted to the ASIC vendor prior to the actual prototyping of an ASIC.

THE IMPERATIVE FOR ASIC SIMULATION

Unlike with printed circuit board design, creating a physical prototype of an ASIC is highly impractical. However, there is a clear need to evaluate the functionality of the ASIC before silicon is produced. Most designers agree that attempting to produce an ASIC without the aid of logic simulation is foolhardy and is likely to lead to extremely high nonrecurring engineering costs (NREs) by producing ''turns'' of the ASIC that do not perform correctly. Logic simulation has grown to meet the need for a ''software prototype,'' performing the same functions as a physical prototype but in a software simulation environment.

The demands on simulation are growing at an ever-increasing rate. In today's world, it is not sufficient to simply verify that the design's logical functionality is correct. In today's modern ASIC, temperature, voltage, and layout are first-order effects that require high simulation accuracy. Designers apply logic simulation and timing analysis tools to allow these factors to be considered early in the design process.

Technology-independent design is adding a new dimension to logic simulation as more and more designers wish to delay the actual choice of technology to later in the design process. For example, a designer may wish to evaluate the design as a 1-μm CMOS device and then as a 0.5-μm CMOS device to determine the cost and performance benefits of each implementation. Delivering this type of capability is a trend in logic simulation.

Vendor-independent design is becoming as important as technology independence. Designers and government agencies are starting to demand the ability to produce designs in a generic manner so that a vendor can be selected on the basis of criteria such as cost, lead time, and second-source availability after the design is fully completed.

Not all simulation environments provide the same levels of capability in these areas. However, the clear trend in logic simulation is toward providing more choices and even more design freedom.

THE COMPONENTS OF LOGIC SIMULATION

There are four components necessary for logic simulation:

- A description of how the cells in the ASIC are connected together
- Models that describe the operation of each cell
- Stimulus to test the design
- Simulation control commands

Logic Description

The logic description is normally supplied to the simulator as the result of schematic capture. Some simulators require a translation step to prepare the schematic capture data for use by the simulator. The translated logic description is sometimes referred to as the netlist. Figure 5-2 is an example of a typical netlist using the venerable TDL format.

Netlist generation is usually accompanied by a need to reduce the design to a single level of hierarchy. This process is referred to as *flattening* or *expanding* the design. More advanced simulators have the ability to directly read the connectivity data created by the schematic capture tool and thus avoid the overhead of netlist creation and expansion.

```
COMPILE $
DIRECTORY MASTER $
MODULE MUX8///DEMOLIB
$
INPUTS
SN1 SN0 S1 S0 B7 B6 B5 B4 B3 B2 B1 B0
A7 A6 A5 A4 A3 A2 A1 A0
$
OUTPUTS
ZN7 ZN6 ZN5 ZN4 ZN3 ZN2 ZN1 ZN0
$
DESCRIPTION
$
LEVEL FUNCTION
$
USE
MUX21LAP///
WIREDAND = WANT(2,1 + TRISTATE SETZZ)
$
DEFINE
ZN7 = (UA7(Z)) $
ZN6 = (UA6(Z)) $
ZN5 = (UA5(Z)) $
ZN4 = (UA4(Z)) $
ZN3 = (UA3(Z)) $
ZN2 = (UA2(Z)) $
ZN1 = (UA1(Z)) $
ZN0 = (UA0(Z)) $
UA6 = MUX21LAP(SN1=SN,S1=S,B6=B,A6=A)$
UA7 = MUX21LAP(SN1=SN,S1=S,B7=B,A7=A)$
UA4 = MUX21LAP(SN1=SN,S1=S,B4=B,A4=A)$
UA5 = MUX21LAP(SN1=SN,S1=S,B5=B,A5=A)$
UA3 = MUX21LAP(SN0=SN,S0=S,B3=B,A3=A)$
UA2 = MUX21LAP(SN0=SN,S0=S,B2=B,A2=A)$
UA1 = MUX21LAP(SN0=SN,S0=S,B1=B,A1=A)$
UA0 = MUX21LAP(SN0=SN,S0=S,B0=B,A0=A)$
END MODULE $
END COMPILE TDL $
COMPILE $
END TDL $
```

FIGURE 5-2. Netlist.

Models

Models are the key to successful ASIC simulation. The extent to which the various models truly represent the actual physical components is the limiting factor in the ability to have confidence in the simulation. Conceptually, models can be thought of as consisting of two components, function and timing. The functional component describes the boolean values that result from applying stimulus to the model's inputs. The timing component of the model describes when the results occur in time.

All simulators have a selection of built-in low-level simulation models known as *simulation primitives*. Primitives usually consist of the simple boolean devices (ANDs, ORs, XORs, etc.) and some minimal selection of sequential devices such as latches and flip-flops. Primitives for RAMs, ROMs, and PLAs are also commonly found. More complex models are created by connecting primitives together on a "sheet" much as any block of logic would be designed. For example, a sheet of simulation primitives might be designed to represent a cell in an ASIC vendor's library. Therefore, a vendor cell in the ASIC design will be represented by sheets of primitives when the design reaches the logic simulator.

The major benefit of modeling with simulation primitives is simulation performance since the logic simulator can be optimized to rapidly evaluate these intrinsic devices. The major disadvantage is a lack of flexibility since only limited parameterization of primitives is generally allowed. This can cause fairly simple ASIC cells to be represented by a great number of simulation primitives, causing a reduction in overall simulation performance and capacity.

Since modeling exclusively with simulation primitives is so restrictive, various additional modeling methods have been developed for today's modern simulator. The ability of a simulator to simultaneously simulate a design with several different types of simulation models is referred to as mixed-model simulation, or sometimes mixed-mode or mixed-level simulation. *Do not confuse mixed-model and mixed-signal simulation.*

Figure 5-3 demonstrates a modeling method consisting of a truth table for the functional description, and Figure 5-4 shows the timing equations to evaluate the timing component.

VHDL is becoming a widely accepted language not only for architectural descriptions but for cell descriptions as well. Figure 5-5 demonstrates how an ASIC cell that mimics the functionality of a 74LS161 might be described by a VHDL description.

A wide variety of additional modeling methods have evolved to support the analysis tools available today. A common modeling method is known as behavioral language modeling (BLM); it permits the binding of C or Pascal models directly to the simulator through dynamic linking. This is a very powerful technique that has the benefit of being extremely flexible. The disadvantage is that it gives designers all the rope they need to hang themselves if they are not careful.

Compiled code models reduce the description of the component to machine code. Models of this type are very efficient, although they nor-

```
##############################################################
##
##   Library:      ACME
##   Technology:   2u CMOS
##   Part:         fdrc
##
##   Description: D flip-flop with rising edge, asynchronous
##                clear
##
##############################################################
##
model fdrc: table =

input d, rn;
edge_sense input cp;
output q, qn;

state_table

     rn,   cp,   d,    q     ::    q,    qn;
####-----------------------------------------
     0,   [??], ?,    ?     ::    0,    1;
     1,   [01], ?,    ?     ::    (d),  !(d);
     1,   [?0], ?,    ?     ::    N,    !(q);
     1,   [1?], ?,    ?     ::    N,    !(q);

end[fdrc: table];
```

FIGURE 5-3. Truth table of a D-type flip-flop.

mally also come with a variety of implementation restrictions that can limit their usefulness.

A number of other VHDL-like hardware description languages are also available from CAE vendors. These languages have the advantage of being tuned for the simulator they support and have the disadvantage of being nonstandard. Some of these languages, such as the Verilog hardware description language, are extremely popular.

Where does a designer obtain simulation models for an ASIC design? As one can well imagine, simulation models can be quite complex, since all possible combinations of input sequence and output response must be determined. Since the integrity of the simulation models is fundamental to the ability to obtain highly accurate simulation results, ASIC vendors almost always provide these models to their customers. It is also likely that only simulations conducted with models approved by an ASIC vendor will be acceptable to that vendor.

Evaluation of an ASIC vendor should include a review of the types of models the vendor provides, covering such items as:

```
###################################################################
##
##   Library:     ACME
##   Technology:  2u CMOS
##   Part:        fdrc
##
##   Description: D flip-flop with rising edge, asynchronous
##                clear
##
###################################################################
##
model fdrc: timing =

declare

#define eq(a)        (der_factor*a)
#define rf_eq(a,b)   (der_factor*(a*inet_load(b)))
#define inet_load(b) (sim_$sum_eval(b,"icap_pin",DRIVEN) \
          + sim_$net_eval(b,"icap_net"))

     table icap nointerp [3] = {
          ["cp"] = .041;
          ["rn"] = .058;
          ["d"] = .154; }

begin

     tP = eq(3.64) on cp(AH) to q(LH);
     tP = eq(3.63) on cp(AH) to q(HL);
     tP = eq(4.44) on cp(AH) to qn(LH);
     tP = eq(5.34) on cp(AH) to qn(LH);

     tP = eq(1.60) on rn(AL) to q(AA);
     tP = eq(2.96) on rn(AL) to qn(AA);

     tR = rf_eq(1.15,"q") on q;
     tF = rf_eq(1.11,"q") on q;
     tR = rf_eq(1.09,"q") on qn;
     tF = rf_eq(1.02,"q") on qn;

#####  Constraint section

     ts = eq(1.00) on d to cp(LH) do {
          sim_$set_state("q", Xs);
          sim_$set_state("qn", Xs);
          sim_$send_msg("Error: Setup violation detected");
          };

end[fdrc:timing];
```

FIGURE 5-4. Timing equations for D-type flip-flop.

```
--  HDL model of a 74LS161-type 4-bit synchronous counter
--  with asynchronous clear

USE mentor_base.ALL;

--  Entity description

ENTITY sn161 IS
     PORT ( clock          :  IN          qsim_state ;
            load           :  IN          qsim_state ;
            clear          :  IN          qsim_state ;
            enable_p       :  IN          qsim_state ;
            enable_t       :  IN          qsim_state ;
            data_a         :  IN          qsim_state ;
            data_b         :  IN          qsim_state ;
            data_c         :  IN          qsim_state ;
            data_d         :  IN          qsim_state ;

            ripple_out     :  OUT         qsim_state ;
            qa             :  INOUT       qsim_state ;
            qb             :  INOUT       qsim_state ;
            qc             :  INOUT       qsim_state ;
            qd             :  INOUT       qsim_state ;

end sn161 ;

ARCHITECTURAL example OF sn161 IS

    FUNCTION bit_to_integer(i3, i2, i1, i0 : qsim_state)
RETURN
            integer IS VARIABLE i : integer :=0 ;
    BEGIN
            IF i3 = '1' THEN i :+ i + 8 ; ENDIF ;
            IF i2 = '1' THEN i :+ i + 4 ; ENDIF ;
            IF i1 = '1' THEN i :+ i + 2 ; ENDIF ;
            IF i0 = '1' THEN i :+ i + 1 ; ENDIF ;
            RETURN i ;
    END bit_to_integer

    PROCEDURE integer_to_bit(i_in : IN integer ;
            i3, i2 ,i1 ,i0 : OUT qsim_state
            VARIABLE i : integer := 0 ;
    BEGIN
        i := i_in;
        i3 := 0;
        i2 := 0;
        i1 := 0;
        i0 := 0;

        IF ((i < 16) and (i >= 8))
              THEN i3 := '1'; i := i - 8 ; END IF;
```

FIGURE 5-5. VHDL description of a 74LS161.

```
        IF ((i <   8) and (i >= 4))
              THEN i2 := '1'; i := i - 4 ; END IF;
        IF ((i <   4) and (i >= 2))
              THEN i1 := '1'; i := i - 2 ; END IF;
        IF ((i <   2) and (i >   0))
              THEN i0 :=                  END IF;

    END integer_to_bit ;

    PROCEDURE shift_out(register_value : IN INTEGER ;
              SIGNAL X3, X2, X1, X0 : OUT QSIM_STATE) IS

        VARIABLE i3, i2, i1, i0 : qsim_state ;
    BEGIN
        integer_to_bit(register_value, i3, i2, i1, i0) ;
        x3 <= i3 ;
        x2 <= i2 ;
        x1 <= i1 ;
        x0 <= i0 ;
    END shift_out;

BEGIN

--  PROCESS (clock'RISING) ;
    PROCESS

        VARIABLE count    :    INTEGER := 0;

    BEGIN
        IF (clear = '0') THEN        -- Clear outputs to 0
          count := 0;
        ELSIF (load = '0') THEN      -- Load Outputs
          count :=
          bit_to_integer(data_d, data_c, data_b, data_a) ;
        ELSIF ((enable_p = '1') AND (enable_t = '1')) THEN
          count := count + 1 ;       -- Count up
          IF count = 16 THEN
               count := 0 ;
          END IF;
        END IF ;

    -- Output values
        shift_out(count, qd, qc, qb, qa) ;

    WAIT ON clock, clear UNTIL ((clock = '1') or
          (clear = '0')) ;
    END PROCESS ;

    ripple_out <= (qa AND qb AND qc AND qd AND enable_t) ;

END example ;
```

FIGURE 5-5. *(continued)*

- The performance of the simulator with the modeling method the vendor has chosen. Generally speaking, primitives run the fastest and high-level models such as VHDL run slower.
- The level of abstraction the vendor has chosen. In many cases a high-level description of a complex cell will run faster than its low-level counterpart. For example, it would take several dozen primitives to model the 74LS161-like cell in Figure 5-5. Even though each individual primitive evaluates very quickly, the cumulative effect of a large number of primitives may make the high-level description a more efficient modeling technique.
- The maintainability of the model library. High-level models are usually easier to maintain since they are more human-readable.

As can be seen, there are many variables at work in the selection of the "best" modeling technology for a specific task, and the ASIC vendor will make a selection based on a best fit of performance, functionality, and maintainability.

Stimulus

The stimulus is developed by the designer in such a way that it increases confidence in the design's ability to perform to its specification. In general, the stimulus represents the 1s and 0s that are applied at specific times to the primary inputs (pads) of the design.

> ***Warning:*** **The designer cannot assume that simulation stimulus correlates to a high degree with the signals that will actually be applied to the ASIC in its circuit board environment. It is very easy to create simulation stimulus that will not actually occur on the board. Detecting these types of situations is generally the responsibility of the developer of the simulation stimulus. It is also very easy to create incomplete stimulus. Fault simulation can be an aid in detecting these situations.**
>
> **In many ways, the stimulus can be considered to be a model of the environment in which the ASIC resides.**

Modern simulators come complete with different methods for entering stimulus. These methods include graphical waveform input and different sorts of stimulus languages. Graphical waveform input is very intuitive but can become tedious for complex input sequences. Stimulus languages are usually more powerful than graphical methods, although they are more difficult to learn. A mix of stimulus development methods is the mark of a good logic simulator.

The simplest method of stimulus entry is the ability to apply boolean values to the primary inputs of the design. This capability is sometimes referred to as list input. In many ways, list input (shown in Figure 5-6) is analogous to assembly language programming. It should be used when precise control over the application of stimulus is required.

```
D CLK CLR PRE
0     1 0 0 1  #  D and PRE =1, CLK and CLR =0
25    1 0 1 1  #  CLR is brought high at 25 ns
50    1 1 1 1  #  CLK is brought high at 50 ns

(Explanation following # is not part of the list input)
```

FIGURE 5-6. List stimulus.

A variant of list input is known as the logfile. Logfile input can be thought of as a compressed version of list input. It is less human-readable than list input, but it is more compact and more easily read by down-stream tools. For example, logfile input is often used as input to test generation translators. Figure 5-7 shows a logfile.

```
  LOGFILE           EXPLANATION (Not part of the logfile)

T 0.0               Current time = 0 ns
D /PRE 20 U         PRE (signal #20) is uninitialized
D /D 18 U           D (signal #18) is uninitialized
D /CLR 21 U         CLR (signal #21) is uninitialized
D /CLK 19 U         CLK (signal #19)is uninitialized
S 18 1              D is high
S 21 0              CLR is low
S 20 1              PRE is high
S 19 0              CLK is low
T 25.0              The current time is now = 25 ns
S 21 1              CLR is brought high
T 50.0              The current time is now = 50 ns
S 19 1              CLK is brought high
```

FIGURE 5-7. Logfile stimulus.

As designs became more complex, higher levels of stimulus were developed. These methods can be either graphical or text-based. Graphical methods are usually better suited to asynchronous types of input where the stimulus is not expected to repeat in regular patterns. Text-based languages often fare better in more synchronous situations where the same stimulus is repeated over and over with small changes. A current trend in stimulus languages is to allow mixing of graphical and textual inputs in a common language. Figure 5-8 is a FORCE stimulus, Figure 5-9 shows an example of a high-level textual language, and Figure 5-10 demonstrates a graphical stimulus.

```
FORCE D 1 -FIXED          D is high at t = 0
FORCE CLR 0 -FIXED        CLR is low at t = 0
FORCE PRE 1 -FIXED        PRE is high at t = 0
FORCE CLK 0 -FIXED        CLK is low at t = 0
RUN 25                    The current time is now = 25 ns
FORCE CLR 1 - FIXED       CLR is brought high
RUN 50                    The current time is now = 50 ns
FORCE 19 1 -FIXED         CLK is brought high
RUN 25                    The current time is now = 75 ns
```

FIGURE 5-8. FORCE stimulus.

```
CIRCUIT exmpl;
/*  This program shows use of the DO and FOR loop
construction */

VECTOR     vec1 = x1 x2 x3 x4 ;
CONST      testpata = '1010's ;
CONST      testpatb = not(testpata);
VAR        I, J;
TIMEDEF    PERIOD = 1200us;

OUTPUT     x5 x6 x7 ;
INPUT      x1 x2 x3 x4 ;
j = 8 ;

FOR I = 1 to J BY 2         /* I takes values of 1,3,5,7  */
{
     vec1 = testpata $      /*  Apply 1010 to vec1 */

     DO I TIMES             /*  DO following 1,3,5,7 times */
     {
          x4 = hi $         /*  Bring x4 to 1 */
          x4 = lo $         /*  Bring x4 to 0 */
     }

     vec1 = testpatb $      /*  Apply 0101 to vec1 */
};
END.
```

FIGURE 5-9. High-level textual stimulus.

FIGURE 5-10. Graphical stimulus. *Courtesy TSSI 8205 S.W. Creekside Dr., Beaverton, OR, 97005.*

Simulation Control

The simulator must also be directed with a series of simulation commands. Examples of these control commands might be:

Setting the timing resolution of the simulator

Selecting signals to be displayed by the simulator

Enabling a logfile to capture simulator results for use by a tester

Instructing the simulator to continue from the end of a previous session

Instructing the simulator to begin execution (run)

Instructing the simulator to stop

These simulation commands may be typed at a command line, be in menus within the simulator, or be entered along with the stimulus. Simulation control commands vary widely among simulators and reflect the features that are incorporated in the particular simulator. The types of commands available in a typical logic simulator are listed below:

BREAK	Defines a breakpoint interrupt
CHECK	Checks for spikes, hazards, or races
CHECK STABILITY	Checks for oscillations
CLOCK PERIOD	Sets time interval for FORCE stimuli
DEFINE BUS	Defines bus for FORCE, MONITOR, TRACE, BREAK, and LIST commands
FORCE	Stimulates signal with a state value
HISTORY	Retains simulation history
ITERATION LIMIT	Limits attempts to generate a stable state
LIST	Displays signal states
LOG	Creates a simulation logfile
READ FORCE LOGFILE	Specifies a file to stimulate a circuit
RESET SIM TME	Resets simulation time to zero
RUN	Starts or continues simulation
SAVE STATE	Writes internal simulator state to disk
SCALE TRACE TIME	Adjusts the trace window time scale
STATUS DRIVER	Transcribes states of all pins connected to a net
TEMPLATE FORCE	Sets default type of FORCE or logfile stimulus
TEMPLATE RUN	Sets default parameters for RUN
TRACE	Displays signal waveforms in trace window
VIEW TIME	Scrolls to specified time in the list/trace window

Displaying Results

Modern simulators offer a wide variety of mechanisms for displaying results from simple lists to graphical displays resembling a logic analyzer. Figure 5-1 (referred to earlier in this chapter) shows a number of these techniques.

INITIALIZATION AND UNKNOWNS

When the power to a printed circuit board is turned on, all signals on the board will stabilize to some logical value: 1, 0, or three-state. Some of these values result from some action on the board's part, such as a power-on reset function. Other signals will have some arbitrary stable value. Over time, the normal dynamics of the circuit board will establish the arbitrarily set signals at different states as required. If the board fails to function properly after power is applied, hardware designers say that the board "has a reset problem" or tends to "lock up."

Software simulators differ significantly in the process that occurs when the design powers up. In logic simulation we have the notion of an *unknown*, or *X*. If a signal is unknown, it could be a 1, a 0, a three-state, or any other logic value the simulator might model. In any computer program, the values of the variables defined by that program are unknown until the program establishes these values. In a logic simulator, signal values are unknown until they are set to a known value by the signal or signals that affect the signal. This process always begins with at least one primary input becoming known. If no primary inputs are known, all signals in the simulator will be unknown.

> ***Rule of Thumb:* Bring all primary inputs to a known state as rapidly as possible. Do not leave primary inputs unknown.**

After the primary inputs to the design have been brought to a known state, many signals will remain at an X. The majority of these can be brought to a known state by exercising the reset lines in the design. However, even at this point there are likely to be a number of signals that are still unknown. Let us look at a simple circuit to see what might cause this situation.

The circuit in Figure 5-11 will always remain at an X state. Designers unfamiliar with simulation concepts will often complain, "But I don't care what value the circuit starts with. Everything will ultimately be in sync." However, the simulator does not understand this argument. In our example, RESET must be changed from ground to a signal that can be toggled from a primary input pin.

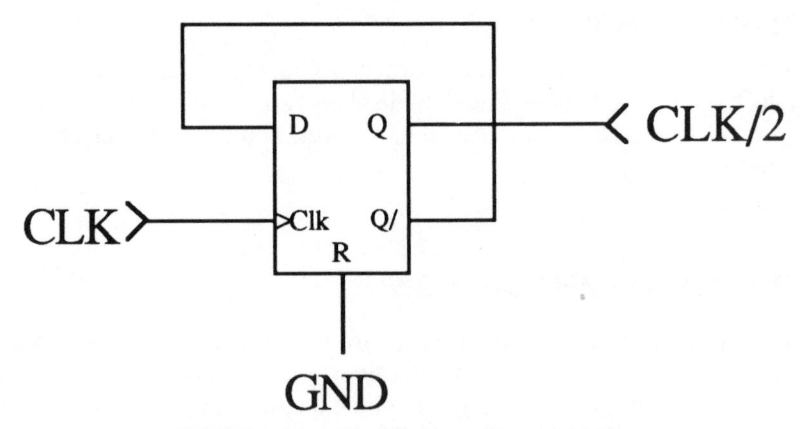

FIGURE 5-11. Circuit that will not initialize.

> *Important:* **Inadequate initialization is probably one of the most troublesome and baffling situations a user is likely to encounter in logic simulation. Be certain the design will initialize through a sequence of transitions on the primary inputs of the design.**

SIMULATION STATES

Simulation states are pairs of logic values and signal strengths. The robustness of the simulation states available in a simulator is an indication of the simulator's ability to accurately model the various situations that occur in a design. This is significant, for example, in correctly evaluating the state of tri-state or wired buses. A more difficult situation to model properly deals with charge storage of MOS transistors, which requires specialized strength information.

Most modern simulators will have at least three logic states (0, 1, and X) and four strengths (strong, resistive, high-impedance, and indeterminate), making a total of twelve available simulation strengths. This number of states is normally adequate for most applications. However, simulators with many more states are available when additional accuracy is required. If simulation at the transistor level is being considered, the designer may very well require a simulator with more than the twelve basic strengths.

As mentioned, an important purpose of simulation states is to correctly evaluate buses. When multiple outputs are driving a single node,

	0Z	XZ	1Z	0R	XR	1R	0I	XI	1I	0S	XS	1S
0Z	0Z											
XZ	XZ	XZ										
1Z	XZ	XZ	1Z									
0R	0R	0R	0R	0R								
XR	XR	XR	XR	XR	XR							
1R	1R	1R	1R	XR	XR	1R						
0I	0I	XI	XI	0I	XI	XI	0I					
XI	XI	XI	XI	XI	XI	XI	XI	XI				
1I	XI	XI	1I	XI	XI	1I	XI	XI	1I			
0S	0S	0S	0S	0S	0S	0S	0S	XS	XS	0S		
XS	XS	XS	XS	XS	XS	XS	XS	XS	XS	XS	XS	
1S	1S	1S	1S	1S	1S	1S	XS	XS	1S	XS	XS	1S

FIGURE 5-12. 12-state node resolution.

the simulator will resolve the value of the node according to a state resolution table. An example of a state resolution table for a typical twelve-state simulator is shown in Figure 5-12.

APPROXIMATIONS AND LIMITATIONS

It is important to realize that proper simulation of an ASIC does not guarantee that the ASIC will work in its intended application. There are numerous factors that can impinge on an ASIC that may not be adequately modeled by the digital simulator. Some of these factors include:

Temperature sensitivity
Voltage sensitivity
Process variation
Loading effects
Magnetic effects
Radiation

As noted previously, a significant trend in logic simulation is to provide mechanisms to take these factors into account. However, no matter how advanced simulators become, they will only be an approximation of reality.

SELECTING THE COMPUTER POWER

Part of a simulation strategy should be to have the proper simulation platform available for each stage of the design process. Low-end workstations should be used for block-level logic verification and debug, where high interactivity is essential. If the size of the ASIC is close to the industry average, it is likely that the designer will be able to select a workstation whose performance is adequate for all design needs. However, if the ASIC is leading the industry average in size, simulation performance may be disappointing. Here the designer should adopt a methodology that permits the use of higher-end machines. This added performance will be invaluable toward the end of the design process, when schedules are tight and simulation performance may prove to be inadequate.

It is difficult to develop rules of thumb for the performance required. Workstation technology is advancing rapidly, and better software techniques continue to provide higher-performance solutions. Fortunately, as workstation performance increases, the relative cost decreases. Disk and memory pricing are trending down as well, but at a slower rate. Additionally, it is likely that we will also see much-improved graphics performance, allowing designers to rapidly display and analyze results.

Having a high-performance, general-purpose workstation as part of the network has proven to be a very successful and cost-effective strategy. This workstation can double as a simulation platform for large runs and as an available resource for other programming tasks.

TRICKS OF THE TRADE

The quality of results produced by a simulator is related to the quality of the models, the accuracy of the netlist, and the designer's skill in creating comprehensive stimulus. It is not unusual to witness a designer peering into a display and exclaiming, "This simulator's wrong! My design doesn't work that way." However, the simulator is seldom wrong. It simply reports. If the results are not what you expect, look for a problem with a model, the stimulus, or the netlist. Don't blame the poor simulator. However, there are a few common tricks that may come in handy.

Initialization Stimulus

Many designers create a specific stimulus to initialize the design. They run this stimulus before adding stimulus to perform a particular piece of design verification. This allows the designer to always work from a known point. This scheme is useful only if the initialization process is short and the design is fairly modular.

Fooling the Simulator

Sometimes this is useful, but, more often than not, forcing the simulator to provide a desired result will ultimately get the designer into trouble. An example might be the addition to a model of a dummy input that forces the model to a certain state when it is toggled. Another example is forcing internal signals to cause the design to initialize. Be very careful. It is possible to mask a design problem or invalidate the use of the stimulus for test by trying this trick unwisely.

Working Around Model Problems

A designer may encounter a situation in which a working model for a particular cell is lacking. Perhaps the designer is creating his or her own library to make use of an internal simulator or hardware accelerator. It may simply be that there is a bug in the library provided by the ASIC vendor. Whatever the case, it is very tempting to simply forge ahead with the simulation. *Don't do it!* The absence of even a single model will mean that the simulation will supply completely erroneous results.

On the other hand, if the designer is certain that the missing cell has no effect on the area to be simulated, he or she may get away with it. However, the designer should be very careful, since the stimulus may give totally different results when the needed model is eventually acquired. In general, it is a much better plan to never begin simulation until a complete set of simulation models is available.

Creating Stimulus with a Model

Everyone who uses a simulator eventually has the idea that since the stimulus is really a model of the environment in which the ASIC resides, why not create a high-level model, such as a VHDL description, that automatically creates the stimulus for the ASIC? Figure 5-13 illustrates this concept.

Conceptually, this sounds like a very elegant solution to the difficult problem of stimulus creation. The pitfall is the complexity of the resulting

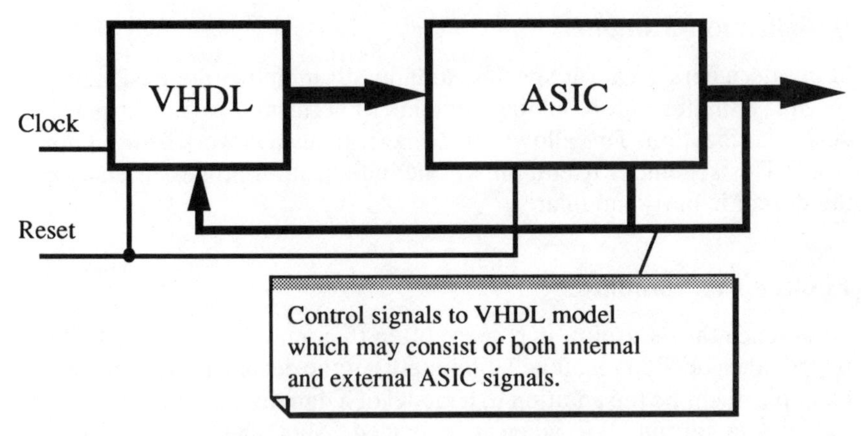

FIGURE 5-13. Creating stimulus with a model.

VHDL model. In many cases, this model becomes so complicated that the designer spends more time maintaining and enhancing the model than working on the ASIC design. Initialization must also be carefully considered or neither the system model nor the ASIC may initialize. Before embarking on this path, the designer should also look at the model debugging tools available with the simulator, since he or she will probably spend a great deal of time tracking problems back through the ASIC into the system model.

The system model is not a solution to be taken lightly and should generally not be attempted until the designer is confident with other modeling techniques. However, under certain circumstances, such as large design teams, this technique can be very effective.

ACCELERATION TECHNIQUES

Logic simulation is, by its very nature, a very compute-intensive task. As designs continue to grow in complexity, designers have turned to acceleration techniques to help them deal with lengthening simulation times.

Acceleration techniques should be considered every time a logic simulation is done. As we have seen, a designer should have a logic simulation plan before even beginning a simulation. Part of this plan must be an analysis of how to maximize simulation performance while providing the required accuracy in the results at each stage of analysis.

Performance Optimization

There are five basic ways to increase the performance of a simulator:

1. **Simulate functional blocks.** The simplest way to improve simulation performance is to simulate small portions of the design before attempting to simulate the entire design. This seems like a simple idea, but it is one that is often overlooked. When the functionalities of each block have been verified, a simulation of all the blocks operating together can be done. There is no reason to waste CPU time simulating the block combination until the individual blocks have been verified and debugged. This technique requires some amount of preplanning since the design must be captured in a hierarchical manner to allow it to be partitioned for simulation.

2. **Reduce the amount of data.** The simulator user has control over his or her own destiny in this regard. Avoid saving excess data to disk. Disk accesses are very time-consuming. Do not simulate beyond the point of interest. It wastes valuable time. Know why a particular simulation is being done. Saving the wrong information, so that a simulation has to be repeated, is worse than saving too much information. Learn to use the simulator effectively.

3. **Raise the level or abstraction.** Higher-level modeling techniques are used to raise the level of abstraction. The key to this technique is to realize that there is a level of abstraction that will provide all the information about the design without providing too much. Consider the 2:1 MUX in Figure 5-14.

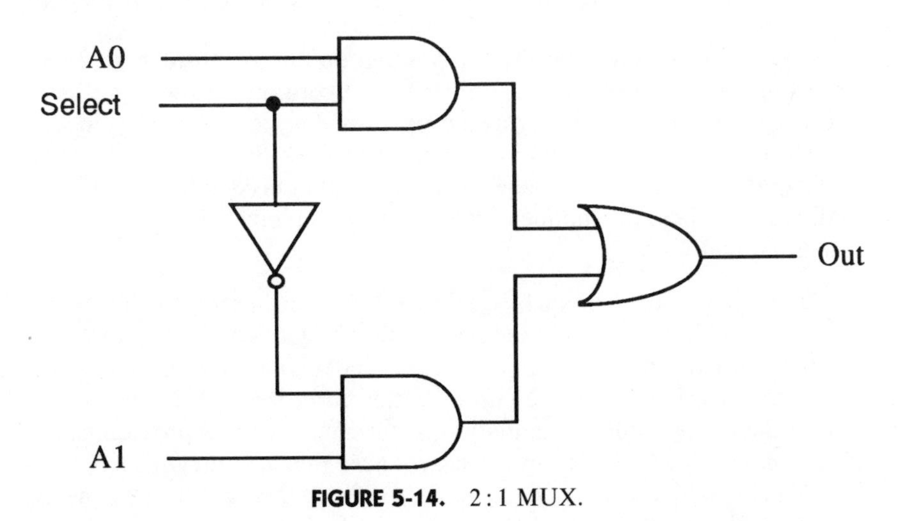

FIGURE 5-14. 2 : 1 MUX.

To the designer, the values of the NAND and INVERTER components of the MUX are of no real interest. The designer is only interested in the inputs and outputs of the MUX. Therefore, if the MUX can be represented to the simulator as a single entity, there will be a four times improvement in performance and memory capacity without loss of information. In reality, the gain in performance and capacity will be somewhat less since the single entity will be larger and more complex. However, simulation performance is always inversely proportional to the number of instances in the design.

Most modern simulators can represent design cells at various levels of abstraction from a switch to an architectural description such as VHDL. The designer's ability to use these methods will vary, to some degree, with the libraries supplied by the ASIC vendor. Occasionally an ASIC vendor will model at a lower level to describe the physics of the cell more accurately. While this is at a lower level than the designer requires, it may be necessary to describe complicated timing relationships accurately.

An effective technique often employed by sophisticated designers is to replace hierarchical blocks with high-level models during the logic verification stage. For example, a counter/decoder circuit may require several dozen individual flip-flops and some amount of discrete logic. However, it could be very simply modeled as a single high-level model during logic design. Toward the end of the design, the gate-level model of the counter/decoder, which was previously verified, would replace the high-level representation. Only a few additional simulation runs are then needed to ensure that the simulation is still giving the correct results.

***4.* Use performance algorithms.** Another option to consider to accelerate simulation is using high-performance algorithms. However, these algorithms obtain their performance by providing less accuracy in the simulation results. This is not necessarily bad since the results only need be accurate enough to allow the logic verification task to be performed. Additionally, these techniques almost always involve constraining the design methodology.

The simulation techniques we have been considering to this point have attempted to accurately describe both the functionality and the timing characteristics of the design. However, under some circumstances not all of this accuracy may be required. In this case, the added accuracy is undesirable since it affects simulation performance without providing any added benefit. The most common case where this accuracy may not be needed is the initial stages of design debug. During this time the designer's goal is simply to verify the correctness of the logic he or she has entered.

The most common performance algorithm makes the assumption that the design will provide the same functional results if all delays are set to one unit. Notice that we are no longer performing timing simulation and that delays between signals can now only be described in terms of "units." It is difficult to generalize the performance gains provided by unit delay simulation since they are dependent on the model's ability to ignore its timing information and the relative efficiency of the simulator in timing versus unit delay mode. However, as a rule of thumb, a unit delay simulator might be three to five times faster than its full timing counterpart. Highly sequential designs without critical timing paths are usually good candidates for unit delay simulation.

A variant of unit delay simulation is zero delay simulation, in which all delay values are set to zero. This technique is not applicable to sequential designs or designs with feedback. However, since it is very fast and works well with nonfeedback combinatorial networks, it may be useful in scan-based designs, which are discussed in Chapter 8.

Another popular technique is called levelized compiled code, or LCC. With LCC the circuit is converted to a single level of hierarchy (levelized) and then converted to machine-executable code (compiled). In the process, all timing information is lost. LCC evaluates all elements in the design for each clock cycle, making it best suited for totally synchronous designs.

5. **Select a higher-performance platform.** If a combination of the techniques we have been discussing is still insufficient to provide the simulation performance required, a higher-performance platform may be needed. However, before selecting such a platform, consider precisely where the performance bottleneck resides.

Quite often the problem will not be a lack of raw CPU horsepower; it may well be lack of RAM. Logic simulation requires a great deal of memory. If the RAM space is too small, the computer is likely to "page." Paging temporarily transfers some of the data in RAM to disk to make room for additional in-memory calculations. When the data placed in temporary disk storage are once again required, they will be brought back into memory. However, to make room, some other data will probably need to be moved to disk. This process is sometimes referred to as "thrashing the disk" and is very much to be avoided. The solution to disk thrashing is to increase the amount of RAM in the computer.

Hardware Acceleration

In essence, a hardware accelerator is a simulator implemented in hardware that provides a high-performance solution to large simulation problems. Most hardware accelerators provide both logic and fault simulation capabilities. The hardware accelerator is normally attached as a server to an engineering workstation that provides the input/output mechanism for the accelerator, as demonstrated in Figure 5-15.

FIGURE 5-15. Hardware accelerator. *Courtesy IKOS Systems, Inc.*

Since a hardware accelerator is a simulator, it has the same basic requirements as any simulator: a netlist, models, stimulus, and control. In many cases a hardware accelerator will be used in conjunction with a software simulator, utilizing a common or translated netlist and stimulus. Since the hardware accelerator is likely to have its own simulation primitives, a mapping between the models used by the software simulator and those required by the hardware accelerator is often necessary. In many cases, the models may be sufficiently incompatible that totally new models will be required for the hardware accelerator.

Since hardware accelerators handle only their own built-in primitives, it may be difficult to map the more difficult models developed for the software simulator. Most hardware accelerators have dealt with this problem by providing a separate execution unit designed for these higher modeling levels. In some cases, the hardware accelerator may use the host computer as the execution unit for high-level models.

When considering a hardware accelerator, make certain the ASIC vendor can supply models to support the selection. Otherwise, you may have a significant modeling chore. If the ASIC vendor does not supply the models for the hardware accelerator, it is likely that they will not accept the simulation results produced by the accelerator. However, the accelerator can still be used for functional and preliminary timing simulation, with the last few runs made on a certified software simulator for submission to the ASIC vendor.

How fast is a hardware accelerator? This is clearly a function of whom you ask and how you ask the question. The actual execution unit of a hardware accelerator can easily be one thousand times faster than the evaluation engine of the software simulator. However, comparing raw performance numbers like these is unimportant since the real concern is how long it takes to obtain results after issuing a run command to the hardware accelerator. What happens when you say "run"?

1. A netlist of the design must be created on the computer (usually a workstation) where the schematic capture files exist.
2. This netlist must then be transferred to the accelerator, usually via a serial link.
3. The stimulus must be translated to the accelerator's syntax.
4. The translated stimulus must then be transferred to the accelerator.
5. The hardware accelerator runs and computes node values. Primitives are evaluated directly within the accelerator; higher-level models are evaluated, when required, by a separate execution unit running at a much slower speed.
6. Results data are transferred from the accelerator to the host computer.
7. These data are interpreted and displayed.

As you can see, a number of translation and communication steps are necessary to perform a complete simulation with a hardware accelerator, and these added steps can reduce the performance of the accelerator by an order of magnitude or more. In addition, the hardware accelerator may need to wait for the high-level model execution unit to complete its task before it can proceed. If this execution unit is very active, the performance of the accelerator can be severely impeded.

> ***Note:* When evaluating a hardware accelerator, be certain to consider the entire simulation process and not simply raw calculation performance.**

Hardware accelerator manufacturers have developed a number of techniques to reduce these bottlenecks. For example, it may be possible to make simple changes to the netlist without going back to schematic capture and creating an entire netlist again. Techniques are then used to import the changes back to the schematic capture tool. These techniques should be used with great care since a divergence between the netlist used by the hardware accelerator and the schematic capture database would ultimately cause a great deal of difficulty. If the netlist changes are not accurately reflected in the schematic submitted to the ASIC vendor, the ASIC will not work correctly and unnecessary NRE charges may result.

DEVELOPING A SUCCESSFUL SIMULATION STRATEGY

Logic simulation should be viewed just as if it were a normal development activity. It should not be approached ad hoc but should occur as part of a reasoned and logical plan. You should also realize that simulation is *part* of design and not a side trip. This means that you should expect to schedule a reasonable amount of debugging time, since things never go as smoothly as planned. You should also have some feeling by now that simulation is not a foolproof process. However, with time and experience, it can be quite rewarding and can greatly enhance your ability as a designer.

The most critical step is to ascertain how simulation models will be acquired. In most cases, simulation models will be provided by the ASIC vendor. However, these models may not be compatible with other analysis tools such as critical path analyzers and hardware accelerators. If one of these tools is to be used, make certain the simulation models work with them. Otherwise it may be necessary to create additional versions of the models, which can result in an added maintenance burden on the project.

As you begin capturing the schematic, it is prudent to consider simulating small sections of the design early on. This will help uncover any unexpected problems in linking the schematic capture process to simulation. When you gain more experience, this step becomes less important. However, experienced designers continue to simulate small sections for a number of reasons, including testing of critical portions of the design and rapid testing of sections of the design that are difficult or time-consuming to manipulate. At each stage of design you should ask yourself, Will I be able to simulate this circuit? Will it initialize? Will the models accurately predict what will actually happen in silicon? Does this part of the design have critical timing dependencies? If you are unsure of the answers to any of these questions, consider simulating the block by itself.

> ***Note:* Do not wait until the entire schematic is captured to find out that the simulator produces unexpected and inaccurate results in a particular area of the design.**

Consider adding testability as each new portion of the ASIC is designed. In many cases, testability can be added without additional real estate, if it is considered early on. However, adding testability late in the design process can be very expensive in terms of both lost development time and die size.

When you are ready to simulate the design "as if it were plugged into the circuit board," it is a good idea to pause and consider the best and most efficient method for applying stimulus. The stimulus must mimic the environment of the ASIC on the circuit board. If the stimulus is rather asynchronous, consider a FORCE-type stimulus language or graphical input. If it is highly synchronous, a structured language combined with a graphical input methodology may be a good approach.

Before writing the first line of stimulus, take a few moments to decide precisely what you want to do. The goal should be to exercise the design in as realistic a manner as possible *in the least number of vectors*. The better you are at balancing vector count with functionality verification, the more efficient the overall simulation effort will be.

The normal first step is to create a set of stimulus vectors that will completely initialize the design. It is generally unwise to leave sections uninitialized unless it is certain that the circuit will eventually bring them to known states. It is almost always better to add some testability to allow the entire design to be initialized with only a few vectors.

A very useful technique is to modularize the stimulus and avoid "reinventing the wheel." For example, if a design creates interrupts that require a specific servicing routine, it may be possible to break these into a group of stimulus vectors that can simply be pasted in every time an interrupt is generated.

As you go about the task of using the simulator to verify the design, continue to ask yourself, "Are the results I am seeing from the simulator reasonable?" If they are not, *stop* and determine where the problem is. It could be an inaccurate model, the stimulus may have placed the design in an invalid state, or any one of a number of situations may have occurred. Do not continue until the situation has been resolved or reconciled satisfactorily.

It is also a good idea to look at signals in areas of the ASIC not currently being tested. Sometimes a stimulus will cause other areas of the design to go into invalid states. This is generally only a problem if nodes in the design go to an X or unknown state. It is usually wise to find out why these situations are occurring and rectify them or they will cause problems later on.

TOP TEN MOST COMMON MISTAKES

We have already identified a number of ills that can affect a logic simulation. Here is a short list that may help the designer do your logic simulation more effectively.

1. **Failure to provide proper initialization of the design.** This can be very frustrating, since the simulation never seems to get off the ground.

2. **Modeling at the wrong level of abstraction.** Having too much detail in a simulation model will dramatically reduce simulation performance. Having too little detail will provide an inaccurate simulation.

3. **Failure to create a simulation plan.** Without a plan, it is not likely that the designer will be able to create reusable stimulus modules. An example of such a module is a set of stimulus that initializes the design. Designers who do not plan find themselves creating and then discarding much of the stimulus they create.

4. **Accepting unreasonable results.** Simulation users are too often prone to accept simulation results at face value. While it is true that simulators seldom produce incorrect results (i.e., the simulator rarely makes a mistake), it is not uncommon for simulation results not to concur with the signals on the actual silicon. In most cases, the problem can be traced to a model or models that fail to adequately handle a particular situation.

5. **Ignoring unknowns.** Unknowns are almost always bad since they generally result from a model's inability to correctly resolve a particular sequence at its input pins. These Xs then begin propagating through the design, eventually forcing the designer to deal with them.

Experienced simulation users have become skillful in using Xs to their benefit in situations where the result may be truly indeterminate for a period of time. However, this is not a technique for use by the weak of heart.

6. **Expecting too much from simulation models.** The ability of a simulation model to accurately predict results is a function of the skill of the modeler and the number of possible input sequences that can affect the model's outputs. The more complex and sequential a cell is, the more difficult it becomes to create a completely defined simulation model.

To demonstrate the situation, consider a D-type flip-flop with Q = 0, RESET and SET = inactive, and D = O. Now suppose CLK is going from a 0 to an X. Some modelers believe that Q should now be set to an X. However, since D = Q = 0, most modelers would leave Q = 0 since it is immaterial whether a CLK has occurred or not. Now what should happen if D goes from a 0 to a 1 while CLK is unknown? Some modelers believe that Q should go unknown, since the X on CLK could be hiding actual transitions on CLK. If this were the case, the 1 on D would be clocked to Q. Other modelers disagree, believing that this is too restrictive since Xs are normally stable (that is, they represent a 0 or a 1; we just do not know which). Imagine the difficulties encountered in accurately modeling a core microprocessor! If you have any doubts about the ability of a model to accurately predict a particular situation, you would be strongly advised to simulate a small block and make sure.

7. **Assuming models work with all analysis tools.** It is very difficult to create a model that is accurate for logic simulation, fault simulation, timing analysis, and other types of analysis tools. Make sure the models are appropriate for the task.

8. **Creating invalid input stimulus.** It is very easy to create stimulus at the primary inputs of the ASIC that will never occur at the pins of the actual ASIC mounted in the circuit board. This may cause the design to behave in unexpected ways. In some cases, the design may lock up or may go unknown. When these situations occur, look carefully to be certain that the sequence cannot actually occur on the ASIC before changing the stimulus. It is not unheard of for these simulation "mistakes" to uncover the need for additional logic in the design.

9. **Not utilizing performance algorithms.** During functional verification, performance algorithms that trade performance for timing accuracy can greatly enhance simulation efficiency.

10. **Failing to adequately consider test.** Many designers would actually consider this to be the single most common mistake new simulation users make. We will consider test in much greater detail in a later chapter, but here is a look ahead at this vital topic.

CONSIDER TEST *NOW*

ASIC designers often mention that developing test vectors is one of the most difficult aspects of designing an ASIC. It is no wonder that test considerations have become a topic of hot discussion in the ASIC community. It is also why we mention test here before considering the topic more fully in a later chapter.

Probably the most common misconception is that test vectors delivered to the ASIC vendor are simply developed from stimulus used to verify blocks of the design. This simply is not so. Testers apply stimulus and measure response at the physical pins (pads) of the ASIC. These pins are sometimes called the primary inputs (PIs) and primary outputs (POs) of the design in simulator parlance. Simulators, on the other hand, can apply stimulus and measure response at any node in the design.

> ***Note:* It is quite easy to develop simulation vectors that cannot be re-created on the tester.**

Therefore, the stimulus used to test the individual blocks cannot be used to develop the test program for the ASIC directly. When designers come to this realization, they are usually tempted to consider concatenating the vectors from the various blocks to create the test program. If the blocks are virtually independent, this may work to some degree. However, as a practical matter, the blocks are usually so interrelated that this proves unsuccessful.

The test program, in many ways, represents a model of the environment in which the ASIC works. The signal interrelationships at the pins of the ASIC are, for this reason, usually complex. Therefore, the most successful way to create a test program is to consider the ramifications of applying simulation vectors only to the primary inputs of the design. Realize that measurements are made only at the primary outputs.

Testability is the efficiency with which a comprehensive test program can be created and executed. Here is a quick checklist that can be used to determine a design's testability.

- **Initialization.** All memory elements must be able to be set to a known state through use of PIs only. It is unacceptable to force initialization with simulation tricks such as forcing internal signals since the tester will be unable to perform these bits of magic.
- **Visibility.** The visibility or observability of the design is a measure of how difficult it is to see the result of the stimulus on the POs. Designs with low visibility are characterized by excessive amounts of deeply buried logic. Monitors for critical internal control functions should be considered to improve the visibility.

- **Controllability.** Controllability is the ability to put a node to a desired value. For example, circuitry that is activated only at the terminal count of a long divider change is not easily controllable. Designs with low-controllability areas should have circuitry added to permit direct control during test.
- **Feedback Loops.** Feedback loops create considerable problems in creating test programs and cause longer than necessary test sequences. Feedback loops can often be broken electrically, permitting a much simpler test program.
- **Counter and Shift Registers.** Long count or shift strings create lengthy simulation times and long test times. Consider providing the ability to control sections of the string independently.

It may be necessary to dedicate pins of the ASIC to test functions that enhance the controllability or visibility of the design. It will be necessary to consider the trade-off of adding dedicated test pins against a higher test development cost if the pins are not used. Making the proper analysis of these trade-offs is critical to producing an ASIC on time with high quality.

Structured techniques such as scan design were developed because of the difficulty of test creation. These techniques generally require more ASIC real estate to be devoted exclusively to test than the ad hoc testability we have been considering so far. They will, however, make test almost automatic. Scan-based design is explained more fully in Chapter 8.

There are a number of additional practical test considerations, including test pattern depth, tester memory limitations, and test pattern load times. These topics are also considered in Chapter 8. Consider test now. Testability helps provide a design that has higher quality at a lower cost.

BOARD-LEVEL SIMULATION

Logic simulation of the entire circuit board is becoming increasingly popular, since it is one of the best techniques for doing board-level analysis very early in the design process. The designer is in the best position to help drive the decision for board-level simulation.

SUMMARY

Much of the material in this chapter represents rules of thumb to aid designers in successful logic simulation. You may find some of them more valuable than others. The important concept is to develop a strategy and a methodology that works for you and allows you to work *with* the simulator and not against it.

To reemphasize a few points:

- Develop a simulation plan. You would never think of starting a design, for example, by placing a NAND gate in the center of a sheet and then considering what to do next. Simulation should be a logical progression of design verification.
- Always watch for and correct unreasonable results.
- Be suspicious of unknowns.
- Make sure the models can support the kinds of stimulus to be applied to them.
- Check to be certain that the models also support the other analysis tools you intend to use. If not, it will be necessary to acquire another set of models for this purpose or inquire into the possibility of extending the ones you have.

REFERENCES

Abramovici, Miron, Melvin Breurer, and Arthur D. Friedman, 1990. *Digital Systems Testing and Testable Design* pp. 39–72, New York: Computer Science Press.

Armstrong, James R., 1989, *Chip-Level Modeling with VHDL,* Englewood Cliffs, NJ: Prentice-Hall.

Breuer, Melvin, and Arthur Friedman, 1976, *Diagnosis & Reliable Design of Digital Systems,* Rockville, Maryland: Computer Science Press, Inc.

Coelho, David R., 1989. *The VHDL Handbook,* Norwell, MA: Kluwer Academic Publishers.

Culbertson, Gary, 1988. Managing the ASIC Design to Test Process. In *1988 International Test Conference Proceedings* pp 649–656, New York: IEEE.

Huber, John P. System Simulation at the Board Level with Electronic Design Automation. Paper read at Electro '88, March 27, 1988, at the World Trade Center, Boston.

Huber, John P. Successful ASIC Verification. Paper read at Electro '89, April 1989, at Jacob K. Javits Convention Center, New York, NY.

Leung, Steven S., and Michael A. Shanblatt, 1989, *ASIC System Design with VHDL: A Paradigm,* Norwell, MA: Kluwer Academic Publishers.

LSI Logic Corporation. 1989. *LSI Logic Primer for ASIC Design*. Milpitas, CA: LSI Logic Corporation.

Mentor Graphics Corporation, 1989, *An Introduction to Digital Simulation,* Beaverton, Oregon: Mentor Graphics Corporation.

Misco, Alexander, 1986, *Digital Logic Testing and Simulation,* New York: John Wiley & Sons.

Turino, John, 1990, *Design to Test,* New York: Van Nostrand Reinhold.

6

Timing

THE IMPORTANCE OF TIMING

It is important to consider the timing needs of an ASIC design to help ensure that the ASIC functions properly in its environment on the circuit board. The component population of a circuit board consists of devices that cover the range of acceptable performance specifications. Some devices will be fast, others will be slow, but all, it is hoped, are within specification. It is important that this collection of disparate parts work together in a manner that meets the overall design specification of the circuit board.

There are a number of factors that can affect the performance of the circuit board. Devices may be at different temperatures, causing some parts to run faster than others. The layout of the board has an effect on timing through the impedance from the traces and the dielectric of the circuit board material. In addition, process variation can cause the timing of each device to vary from the typical specification. All of these effects must be taken into account to ensure that the overall design is reliable and can be manufactured with high quality. As evidence of this fact, almost any manufacturing organization has its "bone pile" of "flaky" circuit boards. A number of these boards suffer from combinations of good components that simply do not "play together."

One of the traditional methods for evaluating timing was simply to test an assortment of prototype devices over a range of environmental conditions. Prototyping should not be relied on as a method of ensuring timing. History is replete with examples of "bad batches" of devices being returned to the ASIC vendor only to have it turn out that there had been a timing problem in the design all along. Prototyping is ultimately a very expensive and ineffective way of ensuring that an ASIC can be designed and produced in a reliable and cost-effective manner.

Over the years, industry pundits have suggested that more than half of all ASICs developed have failed when initially plugged into their circuit-board environments. There could be a number of reasons for this, of which timing problems could certainly be one. It is probably sufficient to note that timing considerations play a significant role in the ability to produce a quality design that works the first time.

BASIC TIMING CONSIDERATIONS

Delay through a circuit is a function of:

The intrinsic delay through each device

The number of loads connected to each net

Temperature

Voltage

Layout (track length and impedance)

Figure 6-1 represents a simple circuit with an AND gate G1 driving two gates G2 and G3. The delay for the output of G1 is given by the equation

$$t_{(d)} = t_{(\mathrm{int})} + K^*L$$

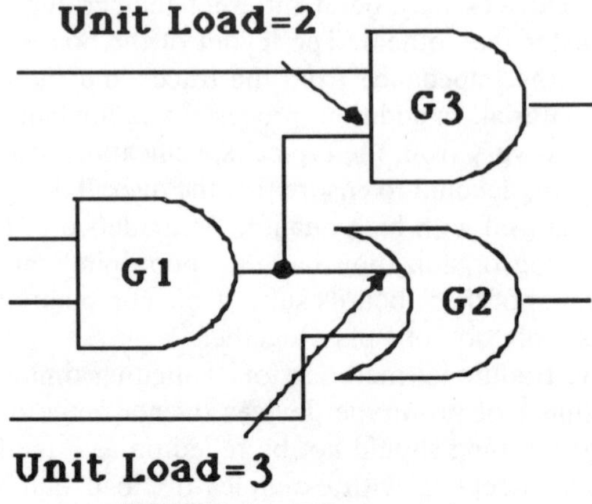

FIGURE 6-1. Delay example.

where:
$t_{(\text{int})}$ = intrinsic delay
K = drive factor (ns/load)
L = Σ equivalent loads = 5 in our example

t(int) is the intrinsic or propagation delay through the device. This number is developed from highly accurate measurements and characterization of the design cell by the ASIC vendor. Intrinsic delay reflects process technology, cell complexity, and cell design techniques.

These timing equations can actually become quite complex. This is particularly true for devices under 1 μm, where loading and environmental effects become first-order in nature. The intrinsic delay through the device is increasingly becoming a second-order effect. In other words, the intrinsic delay through a cell is becoming insignificant when compared with the effects of layout, voltage, and temperature. This is a truly significant development that makes timing considerations all the more important as geometries continue to decrease.

Also, intrinsic delay may not be the same through each path in a device, as demonstrated in Figure 6-2. This phenomenon is referred to as pin-to-pin timing. In the example, the delay from D to Q is 7 ns, whereas the delay from R to Q is only 3 ns. It is vital that any analysis accurately reflect the pin-to-pin timing characteristics of the devices in a design.

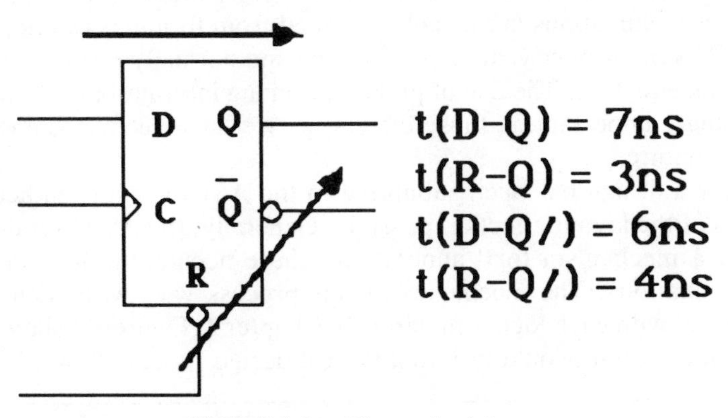

FIGURE 6-2. Pin-to-pin delay.

A related topic is timing constraints: setup and hold constraints, races, and spikes. Setup and hold constraints are associated with sequential cells and require that a signal be stable for a specified period of time at one pin of a cell before a transition occurs at another pin of the cell. A race is said to exist when a circuit can exhibit different behavior depending on variations in timing. Spikes are often the result of races and can normally be considered as small, undesired pulses. Spikes become a concern if they appear at the input of a device with a propagation delay that is less than the length of the spike. We will consider timing constraints in more detail in a moment.

TIMING AND LOGIC SIMULATION

It is possible to allow logic simulation to take the effects we have been describing into account, since almost all modern logic simulators are able to attach timing equations directly to nets and simulation models. These equations calculate a delay that is used by the simulator to more accurately reflect the operation of the circuit.

Timing equations for the particular technology of an ASIC will be supplied by the ASIC vendor. Additionally, the intrinsic delay is a known quantity that is normally part of the simulation model. The logic simulator will probably provide a mechanism to modify the values for the environmental effects (temperature, voltage, and process variation) in some convenient manner. This may be a command line option, or this information may reside as part of a timing file.

The layout information is normally supplied in two forms. Prelayout timing estimations may be available for use before the design has been laid out. These estimations are usually obtained from formulas developed by the ASIC vendor over years of experience and normally reside as part of the timing equation. The use of prelayout timing information is becoming increasingly important as these effects, as we have seen, become first-order in nature.

After a design has been submitted to the ASIC vendor, an accurate analysis of the layout effects is possible. Generally, the ASIC vendor will provide a mechanism for "annotating" these accurate values into the timing equations. The mechanics of this process vary with each ASIC vendor and with each logic simulator. In Chapter 2, Figure 2-1 shows how layout information would fit into a typical design process flow.

***Important:* The ability to accurately analyze the timing requirements of an ASIC is largely determined by the ability of the simulator to comprehend the timing parameters that might affect the design. The timing capabilities of logic simulators vary significantly. Be certain that the logic simulator has the timing capabilities required, and that the data are available and are sufficiently accurate for the current needs.**

It is also worth noting that calculating the timing equations can take a considerable amount of time. While "simply changing the temperature" is conceptually a very simple thing to do, a great deal of time may be necessary to recompute all the equations that are a function of temperature. The time needed to make a change in timing and be ready again to simulate is sometimes referred to as the design iteration time. Timing calculations can take place within the simulator or can exist as a separate process, with the simulator simply reading the final computed values. It is important to analyze the facilities of the simulator and the overall design environment, including the manner in which timing data are provided by the ASIC vendor, when evaluating timing solutions.

HAZARDS

Logic simulators use timing information not only for accurate delay prediction but also to detect timing violations in the design. The types of violations that can be detected include races, spikes, and setup or hold time violations. These hazards sometimes make it impossible for the analysis tool to accurately predict the state of a net as the result of a particular timing event. When this occurs, the state of the net is said to be unknown. In a logic simulator, the net will be set to an X for the duration of the uncertainty. If the net is the output of a sequential device, this X will persist until new data are loaded into the device.

A warning message for the designer is normally displayed when a hazard is detected. Some of these timing errors will actually turn out to be situations that will not affect the operation of the device. For example, a circuit of combinational gates may generate a spike—a small, usually undesired pulse. This spike is probably insignificant if there is no chance for it to be latched by a sequential device. However, if the spike could be latched or could feed to the clock of a device, it should be eliminated.

> ***Important:* All timing violations should be carefully evaluated to determine their effect on the design. Do not ignore timing violation warnings without careful consideration.**

The propagation of spikes through a design during simulation is a particularly important issue, since the presence of spikes on clock lines could have a significant impact on the reliability of the design. There are two prevalent theories about how spikes should be handled by the simulator: inertial delay and transport delay.

The inertial delay technique holds that a spike will not propagate through a device if its duration is less than the propagation delay of the device. Thus, small spikes are "swallowed" as they propagate through the design. This greatly reduces the number of timing violations reported to the designer.

Transport delay suggests that the most accurate approximation for how a spike might propagate can be determined by a pulse that is delayed in time by the propagation delay of the device. Some simulators transport the spike as a logical value, while others transport the spike as an X. Some simulators allow the user to choose between propagating a logical value or an X. As a rule of thumb, most simulators using transport delay will propagate the signal as an X to allow timing errors to clearly manifest themselves. Figure 6-3 illustrates the principles of inertial and transport delay.

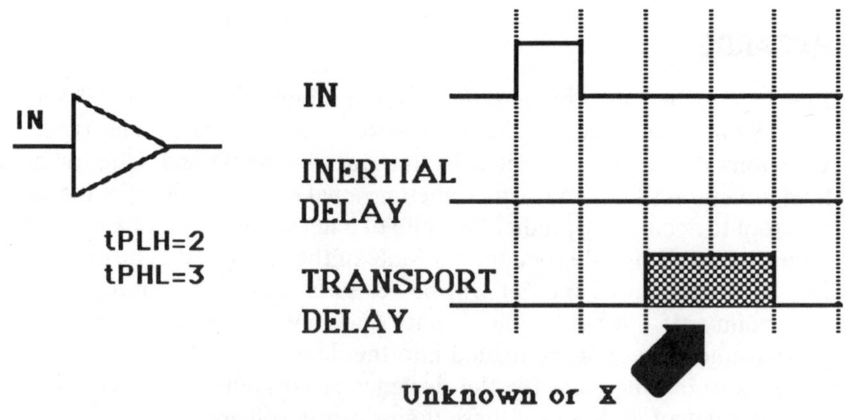

FIGURE 6-3. Inertial and transport delay.

A number of possible permutations of these two spike theories have been adopted over the years. For example, a technique known as unit pulse delay was developed as an alternative to transport delay. Transport delay tends to have a significant impact on the performance of the logic simulator. This is because the generation of the X creates simulation events that must be stored and evaluated at a future time. To combat this problem, a modification of the transport theory suggests that a pulse of unit length is sufficient. This theory works on the assumption that spikes are significant only if they eventually propagate to the clock of a sequential device. If this is true, then the length of the X is not important and a pulse of unit length would be sufficient to cause the output of the sequential device to be forced to an X. The violation has thus been captured and will persist until the device is cleared or loaded with valid data by a valid clock. Figure 6-4 illustrates unit delay spike propagation.

Another spike theory holds that the most conservative approach to describing spike propagation is to cause the output of the device to go unknown immediately when a violation is detected. The unknown will

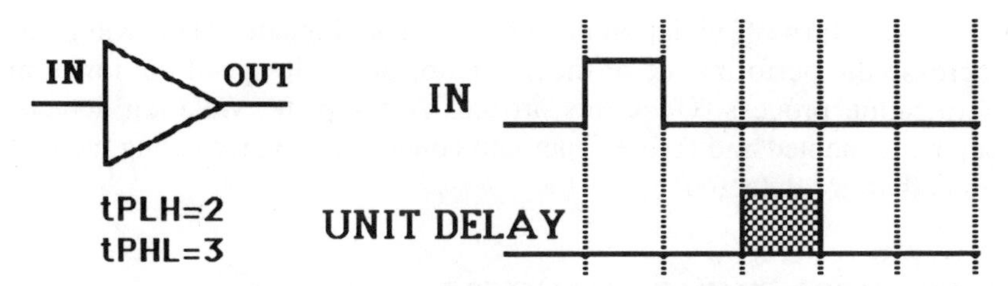

FIGURE 6-4. Unit delay spike propagation.

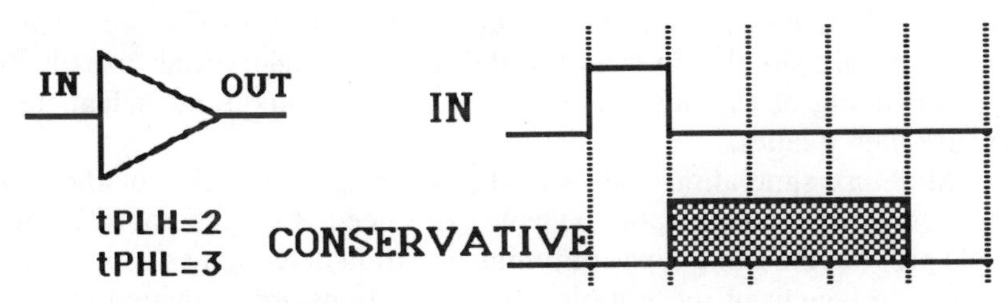

FIGURE 6-5. Conservative spike model.

then persist through the propagation delay of the device. Figure 6-5 illustrates this conservative spike theory.

A trend in logic simulation is to provide the means to mix and match inertial and transport techniques. An example might be to use inertial delay for very small spikes that are much less than the propagation delay of the device and to use transport delay for spikes larger than these small spikes but less than the propagation delay. This technique would help eliminate insignificant spikes while capturing spikes that might affect the device's operation.

No doubt there will be a number of additional spike techniques that will become popular in the future. However, without knowing the detailed physics of the actual device, it is impossible to know precisely how the physical device will respond in the presence of spikes.

> ***Important:* It is likely that the ASIC vendor will have a very strong opinion regarding spike-handling techniques. Be certain that the logic simulator can handle the technique *they* require. Also be certain that the technique they suggest meets *your* requirements for ensuring the quality of the ASIC.**

One last point on logic simulation and timing: The logic simulator should have sufficient flexibility to enable the designer to trade off performance for accuracy. When the designer is simply doing functional verification and is only concerned with debugging the design, it should be possible to turn off violation detection by the simulator. This will greatly increase the performance of the simulator and help speed the functional verification process. Once this process is complete, violation detection can be reenabled and the designer can concentrate on removing the violations that are detected.

SPECIALIZED TIMING ANALYZERS

Experienced designers report that logic simulation by itself is not usually sufficient for analyzing the timing requirements of an ASIC design. This is because logic simulation runs the design at a single speed, usually the typical timing of the device. As we have seen, this is not adequate to ensure high quality.

Min/max simulation and critical path analysis are two of the tools most commonly called upon to answer this need. As a general rule, these tools provide the same types of timing violation detection as a logic simulator. The benefit of these tools is that violations are evaluated over the entire *range* of timing possibilities. In this manner, the designer can gain significantly more confidence that the ASIC meets its complete timing specification.

Min/Max Simulation

Min/max or worst-case simulation is a dynamic timing analysis technique that evolved as an extension of traditional logic simulation. For this reason some logic simulators simply provide min/max capability as a control function. Min/max simulators are also available as separate stand-alone tools.

Min/max simulators operate by simulating the design across the entire range of timing specifications, from minimum to maximum. All timing combinations are analyzed. The region over which signals can change will be displayed, allowing the designer to debug violations. This region is sometimes called the *region of ambiguity*. Figure 6-6 demonstrates the basic principles behind min/max simulation.

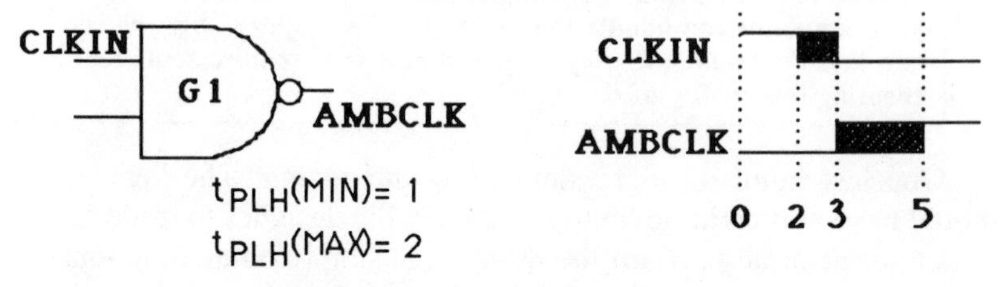

FIGURE 6-6. Min/max simulation.

Notice that CLKIN is shown as a shaded region over which the signal could change from a 1 to a 0. This region is the region of ambiguity for CLKIN. AMBCLK is the inversion of CLKIN when the other input to G1 is at a logic 1. The ambiguity region of AMBCLK is delayed in time from CLKIN in relation to the minimum and maximum specifications for the rise time of G1. Therefore, the region of ambiguity for AMBCLK will occur 1 ns [t_{PLH}(MIN)] after the start of the region of ambiguity of CLKIN. The region concludes at 2 ns [t_{PLH}(MAX)] after the end of the region of ambiguity of CLKIN. Using this technique ensures that the region of ambiguity for AMBCLK represents the entire period over which AMBCLK can change in response to a falling edge of CLKIN.

False hazard reports, where the tool identifies timing violations that do not exist in reality, are a risk associated with any timing analysis tool. In the case of min/max simulation, the ambiguity regions tend to accumulate over the course of a simulation, and this could cause numerous false hazard reports. Min/max simulators use a technique known as *common ambiguity removal* to ensure that only true timing violations are reported. The circuit in Figure 6-7 illustrates an example of common ambiguity removal.

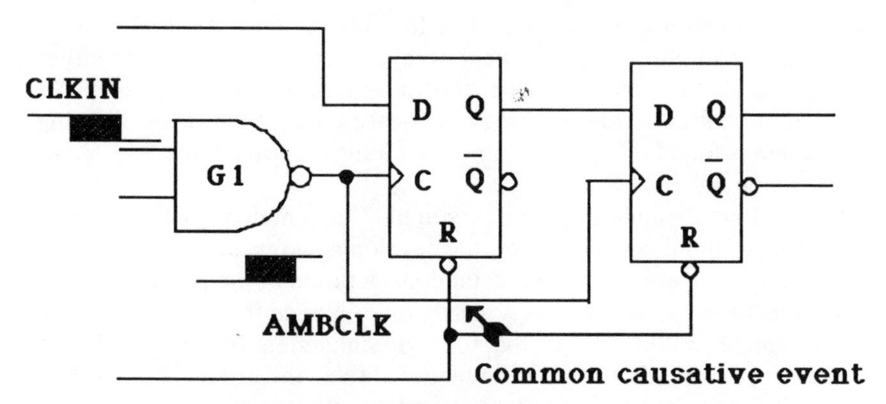

FIGURE 6-7. Common ambiguity removal.

A cursory look at this schematic might indicate that a setup or hold violation could exist at the D input of the second flip-flop as a result of the ambiguity of the clock. This would seem to be the case if the delay from the clock to Q of the first flip-flop plus the hold time of the D input of the second flip-flop is less than the region of ambiguity of AMBCLK. If this were the case, data at the input of the first flip-flop might be clocked to the Q output of the second flip-flop with only a single AMBCLK. This would be indeed unfortunate, since it would prevent us from ever designing a shift register. However, it is obvious that both flip-flops are clocked from precisely the same clock and that the ambiguity of AMBCLK is totally immaterial in determining how the output of the first flip-flop passes data to the second. All that is necessary is for the setup time of the D input of the second flip-flop to be less than the propagation time from D to Q of the first flip-flop.

The common ambiguity removal process recognizes that both flip-flops are clocked from exactly the same signal and that it is not possible for the timing violation to occur. By using a back-tracing algorithm through the circuit, the min/max simulator recognizes the existence of common causative events and properly calculates the regions of ambiguity. False errors are therefore avoided. All modern min/max simulators provide common ambiguity removal.

The min/max simulator we have been describing assumes that there is equal likelihood that any given signal will be fast or slow. For ASIC design, it is possible to refine this analysis because of the close proximity of the transistors and the uniformity of the manufacturing process over the relatively small area of the device. For example, it is reasonable to assume that the process variation between two randomly chosen transistors on the same ASIC will be extremely small. Additionally, while there will be temperature variation across the device, this variation will be relatively small. For these and similar reasons, an ASIC tends to be either all fast or all slow. A technique known as *timing correlation* helps to remove unrealistic hazards during worst-case timing analysis by allowing the designer to establish a more realistic range for minimum and maximum timing values.

A min/max simulator requires stimulus to function, just as does a normal logic simulator. However, the development of a set of stimulus for a min/max simulator requires careful consideration of the possible timing violations to which the design is susceptible. Notice that this stimulus is philosophically different from that for logic simulation. In logic simulation you attempt to verify the logical functionality of the design. In min/max simulation you are trying to create situations that could generate timing violations. This is not an intuitive process and requires a certain amount of experience. In practice, designers start with their functional verification vectors and then add additional vectors to seek out additional violations.

There is no guarantee that the stimulus will flush out all the possible timing violations. The stimulus must create the necessary event to detect timing problems that might be hidden in the design. Fault simulation is not useful for judging the ability of a set of stimulus vectors to detect timing violations. This is because fault simulation tests the ability of the vectors to propagate static faults through the design. These static faults, known as stuck-at faults, have no relation to the timing faults we have been discussing. See Chapter 7 for more information on fault analysis.

Min/max simulation continues to be popular since many timing specifications are state-dependent and min/max simulation is a logical method of investigating timing in the context of a sequence of logic states. However, the disadvantages of min/max simulation are:

- There is no guarantee that all timing violations have been detected, since min/max simulation relies on simulation vectors. If the vectors do not force the circumstance that causes a violation, the designer simply will not know it exists.
- Min/max simulation is substantially slower than logic simulation. A good rule of thumb is that min/max simulators will typically be four to five times slower than logic simulation. In establishing the development schedule for an ASIC, be certain to take this into account if min/max simulation is used.

Critical Path Analysis

Critical path analysis, or CPA, is a type of static timing analysis, meaning that it does not use simulation techniques. Critical path analysis detects timing violations by summing the path delays on datapaths between sequential devices. The critical path analyzer is then able to determine the possible existence of setup or hold violations at the data inputs of the sequential devices. This method is quite similar to the one designers used for years, where waveforms were drawn by hand for all the data signals to determine whether there was a potential for a setup or hold violation.

To a critical path analyzer, a path begins at a source flip-flop and ends at a destination flip-flop. Paths do not go through flip-flops. The exception to this is transparent latches. Of course, paths can also begin at primary inputs and end at primary outputs.

Critical path analyzers report two paths to the destination device. The first path is the one that provides the least amount of available setup time. The second path is the one that provides the least amount of available hold time. In other words, these paths are the ones most likely to produce setup or hold violations and are, therefore, the critical paths. The amount of setup or hold time available is sometimes referred to as slack. Mathematically:

$$\text{Hold slack} = \text{clock period} - \text{hold path time}$$
$$\text{Setup slack} = \text{clock period} - \text{setup path time}$$

A negative amount of slack indicates that a violation could occur. A small amount of slack indicates that there is no violation but that a prudent designer would consider raising the amount of timing slack by redesigning the circuit or reducing the frequency of the clock. In most cases, the correct solution is to redesign the circuit, since the design specifications of the device will probably dictate the clock frequency.

Figure 6-8 is used to demonstrate the principles of critical path analysis.

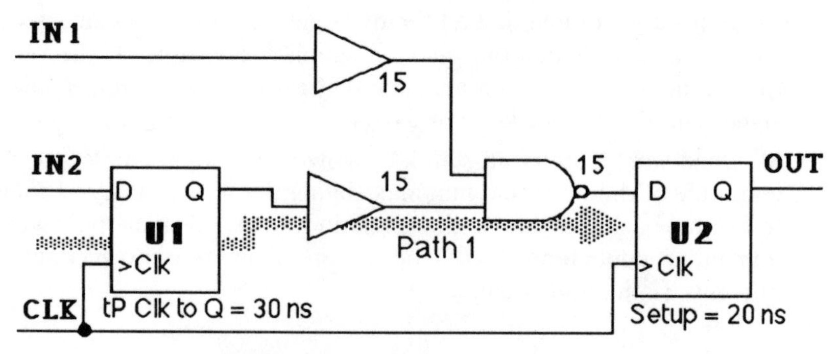

FIGURE 6-8. Critical path analysis.

Looking at path 1, the total delay can be calculated as

U1 tP CLK to Q	30 ns
Buffer	15 ns
NAND	15 ns
U2 setup	20 ns
Total	80 ns

If the period of CLK is greater than 80 ns, there will not be a setup violation for this path. Therefore, a CLK period of 80 ns produces zero slack. Periods greater than 80 ns produce positive slack, and periods less than 80 ns produce negative slack and a setup violation.

Some critical path analyzers utilize only the typical timing for the device. While these tools can be very valuable, they do not detect the possible worst-case timing violations. Worst-case analysis is available in newer generations of critical path analyzers. A popular method for implementing worst-case analysis in a CPA tool is to use min/max techniques on the data and clock lines. To find the worst-case setup time, the minimum delay on the destination register's clock line and the maximum delay on the data line are used. This method brings the clock edge in as early as possible and pushes the data arrival time out as late as possible, creating the worst-case setup time scenario. Worst-case hold time is evaluated by using the maximum delay on the clock and minimum delay on the data.

Critical path analyzers are *not* simulators and do not utilize information about the functionality of the devices in the ASIC. As we have seen, they simply add up path delays and compare the result with the period of the clock at the destination flip-flop. Therefore, the only information the CPA requires is the timing equations for the design. Notice that it is very important that the intrinsic delays be represented as pin-to-pin timing relationships to obtain accurate results from the critical path analyzer.

> ***Important:* The performance of a critical path analyzer is much greater than that available from a simulation-based analysis tool. A good rule of thumb is to expect a critical path analyzer to run between *two and ten times* faster than logic simulation. Critical path analyzers are, therefore, typically *six to forty times* faster than min/max simulation.**

The simple modeling requirements of a critical path analyzer can enable designs that lack simulation models to be accurately evaluated for timing violations. This means that, in addition to its utility for finding violations within the ASIC, critical path analysis is particularly well suited to finding board-level timing violations where functional models may not be available for complex components.

In the discussion of min/max simulation, the concept of common ambiguity removal was described. Critical path analyzers are subject to the same sort of difficulties. Fortunately, all modern critical path analyzers incorporate common ambiguity techniques, as do all modern min/max simulators.

Critical path analysis is not without its own set of difficulties, however. In particular, timing violations that occur as a result of device functionality will generally not be detected. A common example is the race condition demonstrated in Figure 6-9. Notice that if U1 is slower than U2,

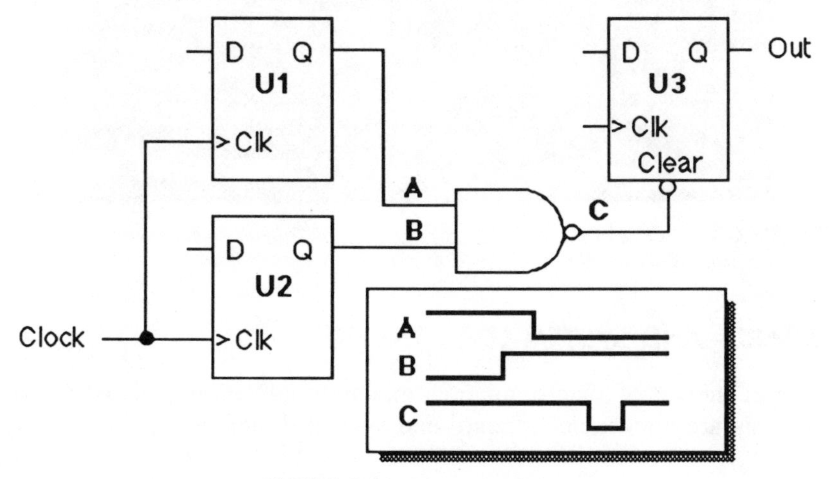

FIGURE 6-9. Race condition.

a spike will occur on signal C that will asynchronously clear U3. However, if U1 is faster than U2, the pulse will not be generated.

Critical path analyzers are also more apt to generate false errors than min/max simulation. Unused paths or unused microinstructions are examples. As a rule of thumb, somewhere between 5 and 10% false hazard warnings can be expected with a critical path analyzer.

Trends in critical path analyzers are toward tools that provide the user with highly graphical feedback on the violations detected in the design. For example, the analyzer shown in Figure 6-10 is able to use path-tracing algorithms and integration with the design database to highlight the critical paths directly on a representation of the schematic. This ease-of-use feature is quite worthwhile, as it reduces the task of eliminating false hazard reports and reduces the need to wade through long timing reports.

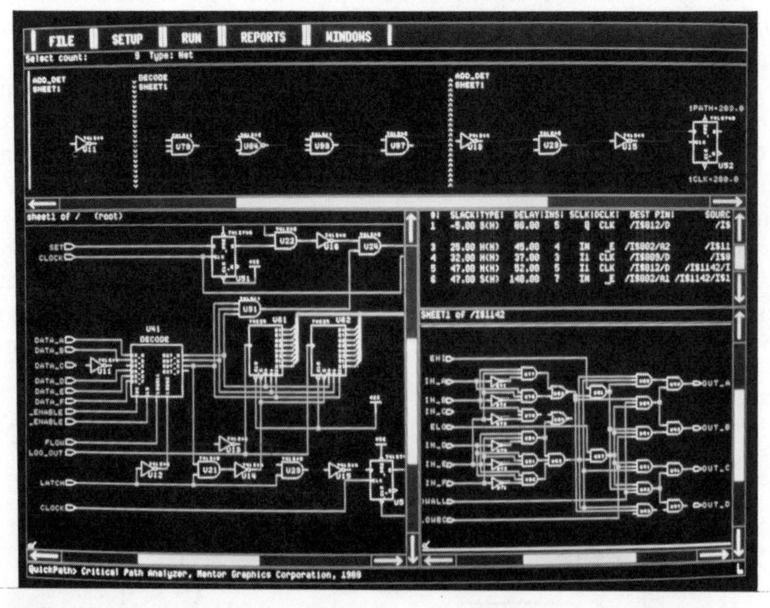

FIGURE 6-10. QuickPath critical path analyzer. *Courtesy Mentor Graphics Corporation, Wilsonville, Oregon.*

BOARD- AND SYSTEM-LEVEL TIMING

The techniques of logic simulation, min/max simulation, and critical path analysis are applicable to board and system design as well as to ASIC design.

Timing analysis at the board and system level is fundamentally the same as analysis at the ASIC level. The key to successfully accomplishing this analysis, however, is the ability to acquire models that can accurately reflect the pin-to-pin timing of the commercially developed devices that are likely to be part of the board design. This may be exceedingly difficult since the timing relationships may be quite complex. For this reason, the pin-to-pin timing for these devices is normally described only at the primary pins of the device and is likely to be an approximation based on values described in the manufacturer's data book. This "timing shell" is normally adequate to obtain valuable information about the timing relationships of the board design.

There are a number of timing problems that are more prevalent at the board and system level than at the ASIC level, such as ringing and crosstalk. A clear trend is for timing analysis tools to offer solutions to these types of problems. We have already discussed the importance of doing both pre- and postlayout analysis at the ASIC level. These techniques are finding widespread acceptance at the board level as well. Board layout tools are becoming increasingly integrated with timing analysis tools to provide this capability. Other types of solutions will probably be offered that will enable specialized routing for high-speed circuitry and automatic termination placement.

A significant need in timing analysis is for timing standards. The ideal situation is for the timing information developed for the ASIC to be completely compatible at the board level. If the board contains multiple ASICs from different ASIC vendors, this can be something of a problem. For example, consider the case of two ASICs from different vendors on a board. If the timing of one ASIC is a function of TEMPERATURE and the other is a function of TEMP, the actual temperature value must be specified in two different ways: as TEMP and as TEMPERATURE. These types of incompatibility in the timing equations make timing analysis at the board level more difficult than it need be. Workstation and ASIC vendors are just now beginning to work closely together to develop standards that will seamlessly enable devices from any manufacturer to be analyzed with the devices from another manufacturer. These incompatibilities should not be a deterrent to performing timing analysis at the board level. However, the designer will need to reconcile these incompatibilities until standards are developed.

SUMMARY

It is vital that the designer adopt a strategy for detecting and correcting timing violations that would affect the operation of the ASIC. This strategy will most likely include the use of at least one of the three general classes of timing tools: logic simulation, min/max simulation, or critical path analysis.

Logic simulation is the most common means by which designers verify timing. The major drawback is that violations are normally detected only for typical device timing. Some logic simulations also offer the ability to simulate with all timing set to either minimum or maximum. This technique, while generating more violations, is usually too pessimistic. You should rely on your logic simulator as a good first approximation of timing analysis, but most designers now agree that logic simulation by itself is not sufficient to fully verify the timing accuracy of an ASIC design.

Min/max simulation is an extension of logic simulation that provides worst-case timing by analyzing the design over all timing combinations. The information relayed to the designer by the min/max simulator is in the form of regions of ambiguity that describe the range over which a signal might change. Common ambiguity removal eliminates false timing errors associated with the tendency of the regions of ambiguity to accumulate as signals propagate through multiple levels of logic. Some of the benefits of min/max simulation include:

- It is a logical and intuitive method.
- Timing specifications are state-dependent, and min/max simulation analyzes timing in the context of a sequence of logic states.

The disadvantages of min/max, however, are:

- There is no guarantee that all timing violations have been detected, since min/max simulation relies on simulation vectors.
- Min/max simulation can be four to five times slower than logic simulation.

A critical path analyzer is a static analysis tool that also performs worst-case timing analysis. This tool adds path delays to the setup or hold delay at a destination flip-flop to determine the presence of setup or hold violations. This technique is identical to the steps a designer would use to verify timing without the aid of a computer-assisted tool. The benefits of a critical path analyzer include:

- All timing violations that are independent of circuit behavior will be identified.
- The speed of a critical path analyzer may be two to ten times that of a logic simulator.
- The critical path analyzer does not require functional simulation models. However, models that describe the pin-to-pin timing of the

device are required. These are, however, generally very easy to create.

- A critical path analyzer does not require simulation vectors.

The limitations of critical path analysis include:

- It does not detect race conditions.
- Errors in the timing model may be difficult to detect and correct. These errors are easier to find with a simulation-based tool. There is no such capability in a critical path analyzer.
- Violations resulting from state-dependent timing will not be detected.
- More false errors will be reported than for min/max or logic simulation.

The timing techniques we have described are applicable at the board level as well as to the design of an ASIC. Ultimately, the only true test of the timing of an ASIC is how well it meets the board timing specification. Prototyping is a dangerous way to ensure board-level timing. Analysis tools should be considered as a way of reducing the probability that a timing problem manifests itself as piles of "dog boards" on the manufacturing floor.

The strategy you ultimately develop will probably utilize a combination of logic simulation and min/max or critical path analysis. Your ability to successfully perform this aspect of the design process will have a significant effect on the overall quality of the ASIC. It is critical that you understand the trade-offs in each method of timing analysis and choose the combination of tools that provides the most effective timing analysis.

REFERENCES

Abramovici, Miron, Melvin Breurer, and Arthur D. Friedman, 1990. *Digital Systems Testing and Testable Design* pp. 74–77, New York: Computer Science Press.

Breuer, Melvin, and Arthur Friedman, 1976. *Diagnosis & Reliable Design of Digital Systems,* Rockville, Maryland: Computer Science Press, Inc.

Bush, Steve. 1987. Automatic Generation of Gate Level Models With Accurate Timing. In *ICCAD-87 Digest of Technical Papers* pp. 52–55, Washington: Computer Society Press of the IEEE.

Hitchcock, Robert. 1982. Timing Verification and the Timing Analysis Program. In *19th Design Automation Conference Proceedings* pp. 594–604. New York: IEEE Publishing Services.

LSI Logic Corporation. 1989. *LSI Logic Primer for ASIC Design*. Milpitas, CA: LSI Logic Corporation.

Martello, Allan R., Stephen P. Levitan, and Donald M. Chiarulli, 1990. Timing Verification Using HDTV. In *27th Design Automation Conference Proceedings* pp. 118–123. New York: IEEE Publishing Services.

Thompson, E. W., S. A. Szygenda, N. Billawala, and R. Pierce, 1974. Timing Analysis for Digital Fault Simulation Using Assignable Delays. In *1974 Design Automation Conference Proceedings*. New York: IEEE Publishing Services.

7

Fault Simulation

WHAT IS FAULT SIMULATION?

Fault simulation is an analysis technique that provides an independent assessment of test vectors before they are used for manufacturing test. The fault coverage is the figure of merit reported by a fault simulator. High fault coverages give an indication of the ability of a set of test vectors to adequately test a design.

Fault simulation is a very compute-intensive task that may be ten to twenty times slower than logic simulation. New software and hardware techniques, which will be discussed later in this chapter, can substantially aid in reducing fault simulation times.

USES OF FAULT SIMULATION

Fault simulation is most often used to gauge the quality of the test vectors submitted to an ASIC vendor by the designer. Imagine the following scenario:

An ASIC is submitted to an ASIC vendor by a designer working for a large company. The ASIC vendor does not require fault simulation but instead requires that the designer certify that the vectors adequately test the ASIC's functionality. Prototypes of the ASIC function satisfactorily, and the design is sent into full production.

Almost a year passes and the ASIC has finally made its way into the end product and into the hands of customers. It is only then discovered that a large number of the ASICs fail to operate in one of its operational modes. Installing a "known good ASIC" corrects the problem. The company contacts the ASIC vendor to determine why the ASICs are of such low quality. The ASIC vendor does a thorough inspection of their process and determines that everything is in order.

Suspecting that the quality of the test vectors may be the culprit, the ASIC vendor runs a fault simulation and determines that the fault coverage is only 70%. A great majority of the undetected faults lie in the area of the design that is failing in the field. The designer has long ago transferred to another project. Much finger pointing then ensues about who is responsible.

This is not a true story, although it certainly might be. The designer and the ASIC vendor jointly have a vested interest in ensuring that the ASIC is tested to high standards of quality.

Fault simulation benefits:

- The designer, by helping to ensure that quality parts are received from the ASIC vendor.
- The ASIC vendor, by ensuring delivery of the quality ASIC that was promised.
- The company, since the cost of finding a fault in test is low compared with the cost of finding it in the field.
- The customer, who demands and deserves high quality.

FAULT SIMULATION IN THE DESIGN PROCESS

Fault simulation normally occurs after logic simulation has been completed and before the design is sent to the ASIC vendor for fabrication. Since fault simulation is normally one of the last steps before the design is submitted to the ASIC vendor, it is often seen as a bottleneck in the design process. It need not be if proper planning is applied up front in the design process.

A continuing trend is to move test-related activities further up in the design process. One important aspect of considering test earlier in the design process is testability. Testability, as we saw in Chapter 5, is the efficiency with which a comprehensive test program can be created and executed. Testability, as applied to fault simulation, means:

- Being able to bring the design to a known state with a small number of vectors. Faults are not detected while the design is at an unknown state.
- Being able to exercise functional parts of the design independently. Exercising portions of the circuit that have already been tested

reduces the efficiency of the test program. Of course, this will always happen to some extent. However, judicious partitioning can improve the efficiency of the test vectors.

- Having an adequate number of observation points.
- Being able to cycle the design through its functions in a minimum number of vectors. Primary culprits in this area are long counter chains that must be continuously exercised to test a portion of the design.
- Being able to readily set signals to a desired state. A design that has this characteristic is said to be highly controllable.

FAULT SIMULATION BASICS

First and foremost, fault analysis approximates failure effects that may exist in the actual ASIC. Please keep this in mind when evaluating the results obtained from a fault simulator.

Fault simulation attempts to mimic, in software, the action of a device tester. Since these testers make measurements only at the primary outputs of the device, faults can be detected only if they can be observed at these primary outputs. Therefore, to detect a fault, the node must be controlled to the opposite state of the fault to be detected and a path must be sensitized or "opened" to a primary output.

Fault simulation generally models two types of faults, stuck-at-1 and stuck-at-0. Fault simulation asks the question, "If the pin of a cell is fixed to a logic level (stuck) and an attempt is made to control it to the opposite state, will the fault be observable at a primary output?" It is important to realize that most fault simulators model stuck pins and not stuck nodes. If fault simulators detected only *nodes* fixed at a logic value, the resulting coverage would be of less use since the fault effects could propagate through many different paths to a primary output. Pin faults, as they are known, are a much more accurate way of implementing the stuck-at fault model.

Deterministic fault simulation is the most popular and relied-upon technique for determining fault coverage. As a general rule, contractual fault coverage requirements are based on the results from a deterministic fault simulator. We will consider the other types of fault simulators later in this chapter when we consider acceleration techniques.

Since deterministic fault simulation utilizes logic simulation techniques, some logic simulation vendors provide fault simulation as a standard feature in their logic simulation. Other vendors provide stand-alone fault simulators separate from the logic simulator.

Deterministic fault simulators answer the question of whether a pin is stuck to a logic value by first running a good circuit simulation. A "fault" is then automatically selected and inserted in a copy of the design in computer memory. Fault simulators use extremely complex algorithms for determining which faults have the potential to be detected by a particular test vector. A simulation is then run on this faulty design and the primary outputs of the device are compared. If they are the same, the fault was not detected. If they are different, then the fault has been detected by the vector. Figure 7-1 shows a screen shot of Mentor Graphics' deterministic fault simulator, QuickFault.

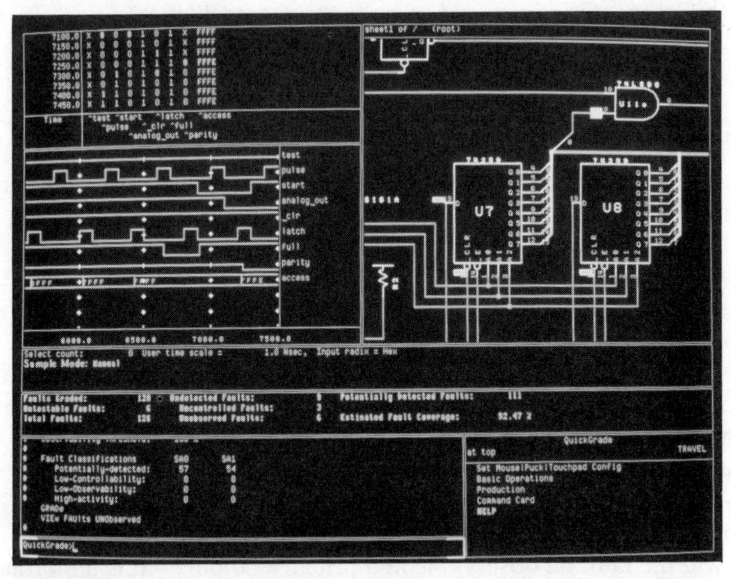

FIGURE 7-1. QuickFault fault simulator. *Courtesy Mentor Graphics Corporation Wilsonville, Oregon.*

One way to visualize how a deterministic fault simulator operates is to consider a pin being disconnected from the other pins on the node and then tied to the logic level of the fault we wish to detect—to ground for stuck-at-0 and to power for stuck-at-1. If the node is then brought to the opposite logic value and this effect propagates to a primary output, the test patterns will have detected the fault.

INPUTS TO THE FAULT SIMULATOR

Just like digital logic simulators, fault simulators require three types of data to function: a circuit interconnect description, cell models, and a set of stimulus vectors. The stimulus patterns are the patterns that will be used by the ATE system to test the ASIC. The fault simulator attempts to emulate the tester environment so that an actual physical device fault is accurately modeled by the simulation. Stimulus for fault simulation is usually a superset of the stimulus developed for logic verification since additional vectors will probably be required to evaluate the design in the presence of manufacturing faults.

OUTPUTS FROM THE FAULT SIMULATOR

In addition to fault-specific information, fault simulators generally provide the same type of output as logic simulators. This allows the designer to trace the values of nets in the design as he or she attempts to establish the conditions that will enable faults to be propagated to primary outputs.

Fault simulators also provide specific fault information, which can include:

- Tabular (ASCII) tables of detected, undetected, and indeterminate faults. (Indeterminate faults are defined in a later section.)
- A graphical annotation of faults on a representation of the schematic diagram of the design. This feature can be a significant aid in improving a low fault coverage.
- A fault dictionary, which is a file containing a list of detected faults for each test vector. The fault dictionary can be used on the manufacturing floor to deduce the likely fault in an ASIC that fails a test program.
- A logfile containing the stimulus and expected results for each test vector. This is the file used to drive the device tester. It should be noted that the logfile represents the results from a good-circuit simulation. Therefore, this file can be generated by either a logic simulator or a fault simulator. It is, however, often more convenient to use the results from the fault simulator since the fault simulator is likely to approximate the tester environment better than the logic simulator.

WHAT IS A GOOD FAULT COVERAGE?

The target fault coverage to shoot for will depend on any contractual obligations for fault coverage, the time available for fault simulation in the design process, and the philosophical bent that the designer and the company have regarding the value of fault simulation in ensuring the quality of the ASIC.

Assuming that contractual obligations do not exist:

- As a practical matter, fault coverage of less than 80% is usually considered inadequate.
- Coverage of 80 to 90% is marginally acceptable but may well be considered inadequate by some organizations.
- Coverage of 90 to 95% is the normal coverage designers tend to use as their target. Coverage in this range is normally quite adequate to provide high confidence in the test program.
- Coverage of greater than 95% is considered exceptional. Obtaining this additional 5% coverage may not be worth the expense. These faults are very difficult to detect and will require a great number of additional test vectors.

ACCELERATION TECHNIQUES

Because fault simulation can be ten to twenty times slower than logic simulation, a number of acceleration techniques have been developed.

LAN Acceleration

LAN acceleration is a technique that allows a deterministic fault simulation to be partitioned across multiple CPUs, typically on multiple workstations connected together over a local area network (LAN). This is done by distributing the fault problem across multiple CPUs. One of the workstations will have the responsibility of distributing the faults to the available processors and then collecting and formatting the results.

A rough estimate of the performance available from LAN acceleration is 0.9 times the number of CPUs used for the analysis. For example, three CPUs will give approximately 2.7 times the performance of a single workstation.

The benefits of LAN acceleration include the ability to utilize existing workstation resources and the ability to obtain added performance by incrementally adding additional or higher-performance workstations.

Toggle Test

Toggle test is not strictly an acceleration technique. However, it is usually considered when conventional fault simulation is too time consuming.

Toggle test is a crude estimate of fault coverage that counts the number of pins that have been controlled to a 1 or a 0. Toggle test does not attempt to determine whether the result of controlling the signal can be observed at a primary output. The "fault coverage" reported by a toggle test will have little meaning since observability is not part of the algorithm. In practice, toggle test fault coverages often differ by as much as 30% from the fault coverage reported by a deterministic fault simulator. In general, statistical analysis, described in a later section, is a much better alternative than toggle testing.

Sampling

Like the toggle test, sampling is a method to obtain rapid fault results. Fault sampling is a technique in which only a small sample of the total number of faults is actually simulated. Analysis is based on probability and statistics theory. The results of a sampling analysis are usually stated as, for example, "95% ± 2% at a confidence level of 90%."

Sampling provides a rapid estimate of fault coverage. However, the list of undetected faults that results is statistically insignificant. This means, for example, that detecting 10% of the faults in the undetected list will *not* necessarily raise the overall fault coverage by 10%. Therefore, while fault sampling is a good method of estimating fault coverage, it does not provide the means to improve the coverage.

Statistical Analysis

Statistical analysis is a technique that also uses probability techniques to provide high-speed fault estimation. In addition to a fault estimate, statistical analysis also provides information on what faults were not detected and why they were not detected.

Statistical analyzers can best be thought of as productivity tools that allow the user to avoid multiple runs through a more time-consuming deterministic fault simulator. In this way, statistical analyzers are complementary to deterministic fault simulators. Figure 7-2 demonstrates how statistical analysis fits into the design process.

Statistical analyzers generally produce an accurate list of faults that were uncontrolled or unobserved by the test vectors. This list of undetected faults is a good basis for developing additional vectors to raise the fault coverage. After all undetected faults have been eliminated from the fault list, the estimated fault coverage is high enough to send the test patterns to a deterministic fault simulator for final analysis. If the deterministic fault number is still too low, a few more deterministic runs will be needed to bring the number up to the required fault coverage. Figure 7-3 is a screen shot of a statistical fault analyzer.

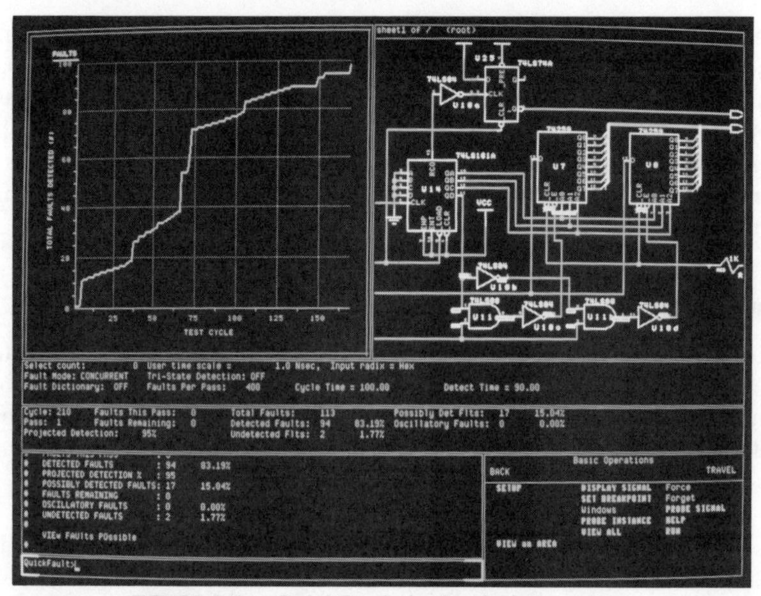

FIGURE 7-2. QuickGrade in the design process.

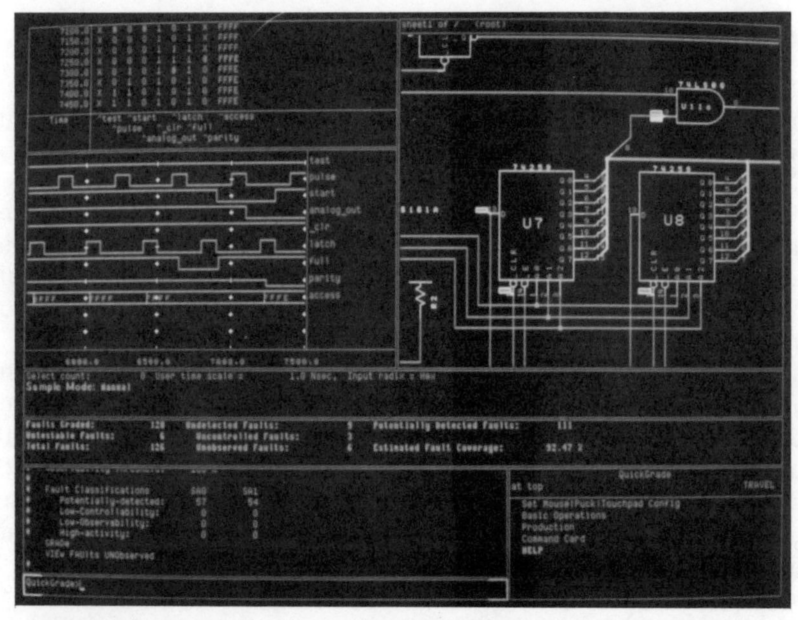

FIGURE 7-3. QuickGrade fault analyzer. *Courtesy Mentor Graphics Corporation Wilsonville, Oregon.*

Hardware Acceleration

As with logic simulation, hardware accelerators can be very useful for fault simulation. They provide a good alternative for generating deterministic fault simulation results in considerably less time than on a general-purpose computing platform. Some hardware accelerators provide fault simulation as a standard option. Other vendors provide fault simulation as an add-on option.

COMMONLY ASKED QUESTIONS

Q. My design verification vectors give me only 70% fault coverage. What's wrong?

A. Probably nothing. Verification vectors often produce a fault coverage of only 40 to 80%. You designed the verification vectors to prove that the design works correctly. The additional 30% is needed to test what happens in the presence of events that do not occur as expected. For example, your verification vectors might always assert a particular input when the design was at a particular state. You also need to test the line unasserted at that state.

Q. OK. I'm up to 90%. How do I get the remaining 10%?

A. Recall that you detect faults when you control signals to the opposite state and then observe the results of these actions at primary outputs. Look for faults that are grouped together in the design. Choose a fault closest to a primary input and create vectors to control this signal to the opposite state. For example, if signal ENABLE is not detected SA0, bring ENABLE to a 1. Now create vectors to propagate the effect of ENABLE at a 1 to a primary output. This takes a bit of practice since it is not an intuitive process. By the way, we chose a fault closest to the primary input since by detecting this fault we may accidentally detect a few of the downstream faults as well.

Q. Well, I've got the coverage up to 98%, but I'm unable to detect the remaining 2%. How come?

A. It may be that the faults are undetectable. One reason might be that you have redundant logic in your design. Redundant logic creates undetectable faults. It is also likely that the fault simulator was unable to exclude faults for signals that cannot be controlled. This often happens with pull-ups, for example. The fault simulator usually knows to delete undetectable faults associated with the side of the pull-up connected to V_{CC}. However, it may not be smart enough to delete the stuck-at-1 fault on the other side of the pull-up. Most fault simulators have a facility for deleting undetectable faults manually before running fault simulation.

> ***Warning:* Be very careful about manually deleting faults, particularly in fulfilling a contractual requirement. Make sure they are truly undetectable. Faults are undetectable if they cannot be controlled to the opposite state and observed at a primary output.**

Q. What is an indeterminate fault?

A. Indeterminate faults are faults resulting when primary outputs change from a known value to an X in the presence of a fault. Indeterminate faults are sometimes referred to as potential or possibly detected faults in some fault simulators. For faults that fall into this category, the deterministic fault simulator cannot ascertain whether the fault was actually detected. Some fault simulators allow indeterminate faults to be counted as detected.

SOME STRAIGHT TALK ABOUT FAULT SIMULATION

Q. Does fault simulation mean automatic test program generation (ATPG)?

A. No. While it is true that ATPG programs often use fault simulation techniques to generate patterns, fault simulation is a distinctly different process from ATPG. Fault simulation provides a measure of effectiveness for a set of test vectors. ATPG programs automatically generate a set of test vectors to a specified level of fault coverage.

Q. Does a fault coverage of 100% guarantee that my design works correctly?

A. No. If it were only so! Fault simulation models a precisely defined class of faults, the stuck-at fault model. While the stuck-at fault model is very useful, there is an infinite number of additional failure modes to which your design could be susceptible. These include timing faults that might become apparent as a result of temperature, voltage, or process variations. Additionally, fault simulation in no way guarantees the functionality of the device.

Q. If I raise the fault coverage from 90% to 100%, will I detect 10% more failing parts?

A. Probably not. This would be true only if faults were randomly occurring, which is an approximation at best.

Q. If I have a fault coverage of 100%, does it mean that all potential shorts have been detected?

A. No. Remember that most fault simulators work with a singularly occurring fault model. There is no guarantee regarding multiple faults. The most common mistake designers make in developing tests is to test a bus by bringing it high and then bringing it low. Shorts across any of the bus lines will go undetected, although the fault coverage will increase as expected. A much better test is to bring each line high one at a time while observing the results at a primary output. This will result in a longer test but will ensure detection of bus shorts. Some newer fault simulators offer the capability to analyze multiple fault situations.

Q. I've made a very small change in the design. Do I need to rerun fault simulation?

A. Probably. It is not unusual for "small changes" to result in substantial reductions in fault coverage. It is very difficult to predict subtle changes that might cause faults that were detected "for free" to now be blocked because of the small change.

> *Murphy's law as applied to fault simulation:* Any change causes the fault coverage to go down.
>
> *Corollary:* The amount of decrease in the fault coverage will be most dramatic toward the end of the design process.

Q. How good is fault simulation for the RAM in my ASIC?

A. Not terribly good, actually. The kinds of faults that affect memory devices must usually be detected through specialized memory tests.

Q. If I have a core microprocessor in my ASIC, can I still do fault simulation?

A. Probably. However, the fault coverage only provides a measure of the effectiveness of detecting faults at the periphery of the microprocessor. For example, there is no guarantee of the processor's ability to execute its various functions. Also, remember that deterministic fault simulation algorithms compare good machine responses with faulty machine responses. The simulation model needs to be totally complete and accurate to guarantee that the faulty machine responses are correct. This is unlikely. In fact, even "full functional models" are unlikely to be able to correctly predict the behavior of complex devices in the presence of an arbitrary fault. The key to fault-simulating complex devices is to ask whether the results will be meaningful and whether they will be reasonable. Fortunately, a pin fault on virtually any pin of a complex part will almost certainly be detected. What fault simulation is unlikely to do is:

1. Give much information about faults internal to the complex device.
2. Tell precisely how a particular fault will manifest itself, since the faulty machine will probably produce Xs for a good part of the simulation.
3. Tell when a particular fault will be observed.

Q. Can I use fault simulation to evaluate precisely what's wrong with devices that fail on the device tester?

A. Yes. This is usually referred to as a fault signature or fault dictionary. However, in practice this technique has some serious limitations. If more than one fault is detected by a vector, which is the usual case, it is difficult to determine precisely which fault is being detected. It may, in fact, be impossible.

CONTRACTUAL OR GOVERNMENT REQUIREMENTS

Contractual requirements are generally stated in simple total fault coverage terms—for example, "The fault coverage of the design must exceed 95%." As we have seen, however, there is a certain amount of ambiguity inherent in the fault number, and further clarification is required. You should determine the answers to the following questions.

Q. Is the fault coverage requirement based on pin faults?

A. The likely answer is yes. All modern fault simulators model pin faults. If nodal fault coverage is all that is required, your task is much easier, although you must find a way to constrain the fault simulator to analyze only nodal faults.

Q. What fault analysis algorithm is to be used?

A. Normally this will be a deterministic fault simulator. However, toggle test or statistical fault analysis can often be used to help minimize the number of deterministic runs that must be made. If you are faulting at the transistor level, you will probably require a specialized fault simulator that is tuned for this application.

Q. How are undetectable faults to be handled?

A. This is a critically important question since it is not unusual to have 5% or more undetectable faults in a design. If your coverage must be 95% and you have 5% undetectable faults, you must really detect 100% of the actual faults. You *must* determine how undetectable faults are to be deleted. Make certain that your fault simulator can handle this process efficiently, and be absolutely certain that the faults you delete are actually undetectable.

Q. How are indeterminate faults to be handled?

A. Recall that these are faults that cause some of the primary outputs of the faulty machine to go to an X. The fault simulator may not be able to ascertain whether a fault was detected. For example, reset lines often appear as indeterminate faults. In practice, the odds are very high that these faults will actually be detected, so indeterminate faults will be counted as detected. You should make certain that either you can count indeterminates as detected *or* you can count indeterminates as undetectable.

Q. What is the appropriate fault level?

A. You must be certain you are faulting with models at the proper level of abstraction. Fault simulation at the transistor level, for example, is much more difficult than fault simulation at the gate level and is often not required. On the other hand, fault simulation of a VHDL description of a block of logic may not be acceptable because of lack of accuracy.

Your simulation models must be able to support faulting at the level you desire. For ASICs this is normally at the design cell boundary. However, some contracts may require you to fault at the transistor level. In this case, you would need to be certain that there is a transistor-level model available for each of the design cells. Be aware, however, that fault simulation at this level can be *very* time consuming. Additionally, the simple stuck-at model may be inappropriate for the types of faults that can occur at this level.

LIMITATIONS OF FAULT SIMULATION

We have stated that fault simulation is only an approximation. The most significant limitation is that fault simulation is based on a singularly occurring fault model. In other words, fault simulators do not attempt to model multiply occurring faults. The most common example of a multiply occurring fault is a short circuit. In the extreme case, if two nodes are always at the same state when the nodes are sensitized to primary outputs, the test will pass even if the nodes are shorted together. The most common mistake in this area is to detect faults on bus lines by first bringing the bus high and observing the effect at primary outputs, then bringing the bus low and observing the effect at primary outputs. If any of the lines of the bus are shorted together, the tester will not detect this defect even if all the faults on the bus were identified as detected.

Fault simulation does not address the variety of other possible faults that could infect an ASIC. Temperature, voltage, and process variation are examples of environmental dependencies that could cause an ASIC to fail to operate. Timing analysis is effective in ensuring reliability in these areas.

Fault simulation in no way guarantees that the functionality of the design is correct. A miscue in the design specification can spell doom for the ASIC when it is installed on the printed circuit board. Board-level simulation has been shown to be effective in verifying the ASIC's ability to function in the board environment.

High-impedance faults have historically been a problem to detect with any type of tester. This is simply due to the fact that faults are detected only at the primary outputs of the device. If the impedance is high enough, the device may pass the test. These devices usually fail to perform at rated speed or suffer early life failures.

Unfortunately, a high fault simulation coverage does not guarantee against the entire range of ills that can affect an ASIC. However, even though a relatively simple fault model is used, experience has shown that a fault coverage of greater than 90% allows detection of a great many of the common ASIC failure modes.

SUMMARY

When used appropriately, fault simulation can be an invaluable aid in creating high-quality test programs. However, do not extrapolate fault detection results unwisely or inappropriately. Fault simulation results are an approximation of the kinds of manufacturing defects that can infect an ASIC. They must be evaluated in the light of the stuck-at fault model.

As with all phases of design, do not leave test to the last step in the design process. Use design-for-testability techniques and concurrent engineering tools such as statistical fault simulators. Use deterministic fault simulators wisely and judiciously, since they are generally very CPU-intensive. Consider a hardware accelerator or LAN acceleration when higher-performance fault simulation solutions are needed.

And remember: Quality cannot be tested into a design. It must be designed in from the start.

REFERENCES

Abramovici, Miron, Melvin Breurer, and Arthur D. Friedman. 1990. *Digital Systems Testing and Testable Design* pp. 131–180, New York: Computer Science Press.

Butler, Kenneth M., and M. Ray Mercer. 1990. The Influences of Fault Type and Topology on Fault Model Performance and the Implications to Test and Testable Design. In *27th Design Automation Conference Proceedings* pp 673–678. New York: IEEE Publishing Services

Fujiwara, Hideo. 1985. *Logic Testing and Design for Reliability* pp 84–102. Cambridge, MA: MIT Press

Giramma, David, Stephen Demba, Ernst Ulrich, Karen Panetta. 1988. Experiences with Concurrent Fault Simulation of Diagnostic Programs. In *1988 International Test Conference Proceedings* pp 877–887. New York: IEEE Publishing Services

Jain, Sunil K., and Vishwani Agrawal, 1984. STAFAN: An Alternative to Fault Simulation. In *1984 Design Automation Conference Proceedings*. New York: IEEE Publishing Services

Mai, Wei-Wei, and Michael D. Ciletti, 1990. A Variable Observation Time Method For Testing Delay Faults. In *27th Design Automation Conference Proceedings* pp 732–735. New York: IEEE Publishing Services

Remeis, Paul. 1990. Applying Fast Fault Grading Techniques Benefits Design and Test Engineering. In *1990 Test Engineering Conference Proceedings* pp 105–118. Boston: Miller Freeman Expositions

Turino, Jon. 1990. *Design to Test*. New York: Van Nostrand Reinhold

Ward, P.C., and J.R. Armstrong. 1990. Behavioral Fault Simulation in VHDL. In *27th Design Automation Conference Proceedings* pp 587–593. New York: IEEE Publishing Services

8

Test

A STORY—PART 1

It was still early and you hadn't even had your first cup of coffee. Too early for silly questions. The young engineer in front of you couldn't be that naive. Don't they teach them any practical horse sense in school anymore?

It should have been an easy assignment even for a neophyte: make scope tracings of each pin for the first 100 ms following reset. Exactly now, what was it he said?

"The ASIC you wanted the scope tracings for seems to have died!"

"There's only two of the new revision in the whole world," you mutter to yourself. "If he's burned up one of them, there's gonna be one less new hotshot engineer!"

"Exactly . . . what did you do?" you ask.

"I wanted to be able to make good, clean connections to the scope. I couldn't do that with the ASIC mounted into the board. I made up a cable so I could get to all the leads and plugged the ASIC into a socket I soldered to the end of the cable. That's when it quit working. The cable is about a foot long. Is that OK?"

Egad. "No, it's not OK." Where do they get these guys? "Look, I went to a lot of effort to minimize the trace lengths on the PC board to achieve the maximum performance from the ASIC. We were particularly careful of grounding. You simply can't take something that runs at 50 MHz and treat it like it's running at 1 MHz. Everything acts like an antenna when you get to these sorts of frequencies. Everything has a complex impedance, and magnetic effects such as crosstalk can be a real problem if you're not careful."

"Oh," your young apprentice sputters. "Makes sense."

"Well, go back to the lab and find a way to make the measurements on the board. And, by the way, make sure you've grounded the probe to maintain the signal's integrity." Oh well, it is fun to teach the new engineers about the finer points of design, but at times it can sure be frustrating.

A STORY—PART 2

"There's nothing like a meatball sub and peppers to get one ready for an afternoon meeting!" you think. The test guy was really insistent. "Hi, Jim, what can I do for you?" you ask as the test engineer strides up to your desk.

"Well, I have a problem trying to make your ASIC run on my tester," he starts. "We may have to reduce the DC test speed below your design spec to get the test to run reliably."

"That's unacceptable! I'm pushing the technology of the ASIC pretty hard, so it's critical that we test at full system speed."

"Look," he says, "there's a finite amount of wire connecting your ASIC to my tester. I've made them as short as possible, but everything acts like an antenna at this frequency. And, you know, my tester simply doesn't have the same impedances at its test pins that your ASIC sees in the board. I'm also having trouble getting good-looking signals since the grounding seems a bit touchy."

"Oh. That makes sense."

A STORY—EPILOGUE

"Well, Jim, we didn't make 50 MHz for the whole test, but I feel pretty good that we were able to make the critical speed measurement with that special AC test you devised."

"Yeah. Give me a call when you start your next ASIC and we can discuss how to go about testing it. By the way, the new kid looks pretty sharp."

"Yeah. He's got a few things to learn, but I think he'll do all right."

THE MORAL OF OUR STORY

An ASIC and the test system connected to it are every bit as much a system as the ASIC plugged into the printed circuit board. The specifications for the interface between the ASIC and the tester must be designed with the same care used in communicating the requirements to the PC board designer.

Since it is not possible to control all the aspects of the tester, it is likely that there will be some amount of attenuation and distortion of the signals in connecting the ASIC to the tester. See Figure 8-1 for the definition of these terms. It is vital that the ASIC-tester interface be tuned to

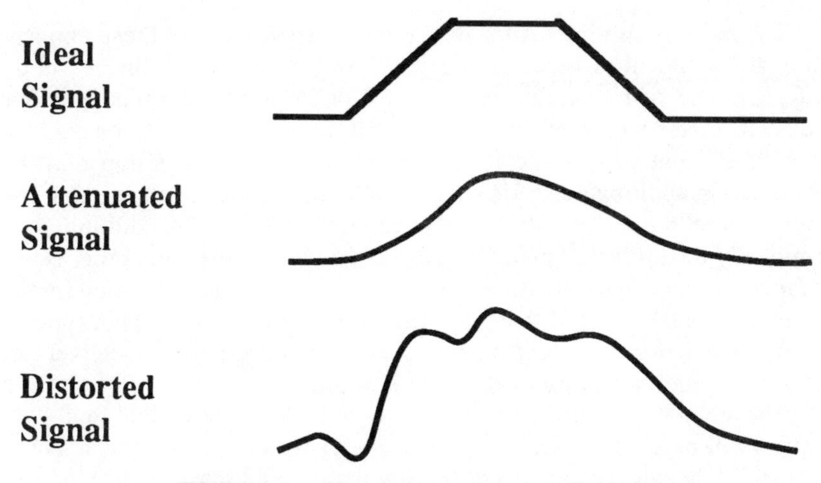

FIGURE 8-1. Signal attenuation and distortion.

minimize this signal degradation. GaAs ASICs, for example, are particularly sensitive to ringing on clock lines since transients could be detected as multiple clocks because of the high switching speeds of these devices. It is very important that signal integrity be maintained for these signals.

Experienced designers take test into account early in the design cycle and investigate how a design is likely to be tested. Realizing that an ASIC must work simultaneously in two diverse environments, the tester and the board, can help create a testable design that moves easily into manufacturing.

TEST AND ASIC DESIGN

The past few years have seen ASIC designers becoming increasingly involved in the development of the test programs used to verify the quality of the finished design. This is clearly a healthy trend, since the designer is singularly in the best position to design a test capable of identifying the subtle but critical failures that might inhabit a defective device.

The designer's contact with the world of test is likely to occur in three ways:

- Providing a test program to the ASIC vendor.
- Interfacing with a test or quality assurance group within his or her own organization.
- Using an ASIC verification tester.

The basic requirements for developing tests for any of these applications are identical. However, the goals and motivations can be subtly different. The ASIC vendor uses the test program to determine whether the ASIC meets its specification. The ASIC vendor expects to be paid for each ASIC that passes the test. The quality assurance group may be interested in auditing the ASICs supplied by the vendor to ensure that the ASICs were properly tested at the vendor's facility. In addition, this group may be interested in investigating early life failures and other types of failure effects analysis. ASIC verification testers tend to be used by the designer to exercise and validate ASICs from the vendor. This type of tester is particularly useful if there is a significant lag between delivery of the ASIC and availability of the circuit board.

The simulation vectors developed for logic verification and fault simulation will most likely serve as the basis for these test programs. These vectors, along with the results of the simulation, will serve as the stimulus and expected response of the ASIC. Output from the simulator is fed to a translator for conversion to the proper tester syntax. These translators are often supplied by the ASIC vendor or tester manufacturer. The simulator vendor may also make available translators to a number of popular testers. Additionally, independent test consulting and software development companies are sources of translators.

The development of a test plan is critical to the successful development of a test program. The goals and objectives of the test should be clearly understood and the designer's role in the process should be clearly defined. The test plan must work within the constraints of the available test hardware and should describe the limitations of the test.

Production testing of an ASIC is normally divided into two parts: DC test and AC test. The goal of DC testing is to check whether the device performs its intended functionality. Certain DC parameters such as power consumption, DC leakage currents (IDD), and output voltage levels (VOL, VOH) may also be performed at this stage. The AC tests are designed to help verify that the device can run "at speed." Tests for propagation delays, maximum frequency, and rise/fall time measurements are the types of tests that are typically performed at this stage.

DC TEST METHODOLOGY

All ASIC testers are used for DC test work in a similar basic manner. The ASIC plugs into a test fixture that is connected to the electronics of the tester. The ASIC is usually referred to as the DUT (device under test) or UUT (unit under test) when it is plugged into the test fixture. The tester is able to either drive signals to the DUT or monitor signals being driven from the DUT. Modern testers can switch between driving and sensing on the fly.

The test system will have an operating system or test executive to control test operation. This executive allows a skilled test engineer to harness the power of the tester to apply signals at precise times and make reliable and repeatable measurements. ASIC verification testers tend to have greatly simplified test executives, allowing simulation vectors to be easily converted to a usable test program. However, these programs tend to be less aggressive in their goals than those developed for larger, more sophisticated systems.

One of the functions of the test executive is to load the translated simulation results into high-speed memory in the test system. Running the test involves incrementing through this memory and applying stimulus to the ASIC or making measurements as appropriate. This is straightforward for ASIC pins that are simple inputs or outputs. Bidirectional pins carry the additional requirement that the simulation results convey not only the state of the pin but also its direction. This is necessary if the tester is to know when to switch from driving to sensing.

The time between applying the data in consecutive memory locations to the DUT is referred to as the *test cycle time* or *test period*. Test cycle times are controlled by a programmable test cycle clock within the tester. Individual tester pins can be programmed to change at programmable time intervals following the start of the test cycle.

The number of test cycles that can be run in a single test program is limited by the amount of high-speed memory. If more test cycles must be run than the memory will permit, the tester must stop and reload memory with the next group of test cycles. Each of these individual test programs is referred to as a burst. Bursts reduce the efficiency of the tester and should be kept to a minimum. Additionally, since the test halts while new data are loaded into high-speed memory, any dynamic data will be lost unless special steps are taken to refresh this information. While this is not normally a problem for ASIC devices, it can be quite important for full custom designs.

The amount of high-speed memory in an ASIC tester is quite limited even in today's modern machines. A size of 4096 locations per pin is still common, with newer testers offering up to 64K per pin. If the target tester has 4096 locations per pin and the design is to be tested at 50 MHz, the total test time per burst is equal to:

$$\frac{4096/(50 \times 10^6)}{2} = 41\ \mu\text{s per burst}$$

Notice that two locations in memory are required to create one test clock. Using two memory locations to generate one test clock is very inefficient. To maximize the use of precious memory, testers utilize the concept of timesets. Typically there are three types of timesets: NRZ, R1, and RZ.

Most tester pins will be defined as non-return zero (NRZ). These pins simply reflect the value stored in high-speed RAM to the pin of the DUT. For example, a 1 in RAM will cause a 1 to appear at the pin of the ASIC. All the examples we have used so far are of type NRZ.

Return one (R1) and return zero (RZ) permit a tester pin to create a pulse or pulses during a single test cycle. This greatly improves the efficiency of the test program by allowing more tests to be run in a single burst. However, simulators generally do not know how the tester will ultimately be programmed, so considerable editing may be necessary in the translated program to impose these timesets on the test program. Additionally, testers do not allow timesets to be reprogrammed during a burst. This means that the stimulus must be periodic during the burst to allow one of these test sets to be used. While the R1 and RZ test sets allow the tester to be more efficient, they can impose considerable restrictions on the logic simulation. Consider using R1 or RZ for clocks and NRZ for data.

Output sensors in the tester make their measurements at the command of a compare strobe slaved to the test cycle clock. The compare strobe is a one-shot event that captures the state of the signal and compares it to the value in high-speed memory. If there is a miscompare, the pin of the ASIC fails for that test step. Outputs and bidirectional pins must also be assigned to a timeset. Outputs are almost always assigned to NRZ, and bidirectionals will always be assigned as NRZ.

Figure 8-2 summarizes the three basic timeset components.

The selection of the compare strobe time is crucial to successful test development. The time between the beginning of the test cycle and the compare strobe is referred to as the propagation, or settling, time. The propagation time must take into account propagation from the drivers of the tester through the fixture to the input of the ASIC. This time also includes the time to propagate the signals through the ASIC and the time to propagate from the ASIC's outputs through the fixture to the comparators in the tester.

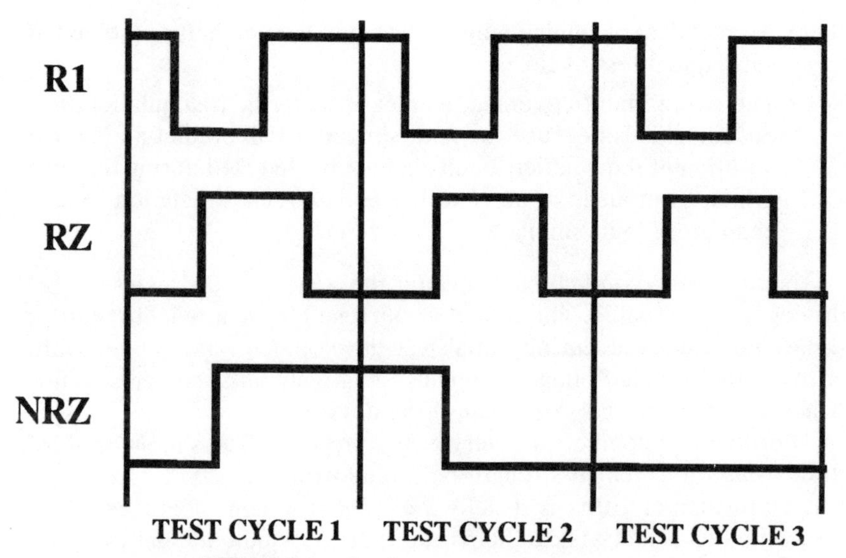

FIGURE 8-2. Basic timeset components.

> ***Important:*** **Testers do not make subjective judgments. A single miscompare is sufficient to indict the ASIC as defective. Many a good component has wound up in the scrap bin as the result of a test program with a compare strobe that has been set too tight.**

It is wise to allow as much time between the start of the test cycle and the compare strobe as possible. The longer the signal is allowed to settle, the more reliable the test will be. This must be balanced against the need to test the ASIC as close to its operating specifications as possible. It is this situation that forces many test programs to operate at less than the actually intended operating speed. Fortunately, when coupled with AC testing, this test methodology has proven to be extremely effective.

There are a number of common mistakes that design engineers unfamiliar with test considerations are likely to make. These include conflicts, floating inputs, and initialization problems. A conflict is said to exist when two signals simultaneously drive a bidirectional bus. The most common example of a conflict occurs when both the bidirectional pins of an ASIC and the pins of the tester are configured as outputs. A more insidious conflict can occur if the simulation vectors cause two drivers within the ASIC to drive a bus simultaneously. Conflicts must be avoided since:

- Damage to the ASIC could result.
- Noise can be generated in the power or ground lines, making the test unreliable.

- A great deal of ringing may occur when the conflict is resolved in a subsequent test cycle.
- The overall fault coverage, as measured by fault simulation, is reduced. This is because the fault simulator will predict an X for the duration of the conflict. Faults cannot be detected during the time a primary output is at an X in the good-circuit simulation. See the chapter on fault simulation for more details.

Notice that a conflict exists even if the ASIC and the tester are both driving the same value. This situation can result in an unreliable test since it may go unnoticed during initial test development and debug. Differences in the driving voltages can cause relatively large currents to flow, which can generate noise or heat in the device.

Fortunately, conflicts are very easy to resolve. The simulator should force a high impedance strength before allowing the ASIC to drive the net. Unfortunately, this is usually very much a manual process. High-level stimulus languages can help the situation by automatically applying the high-impedance state.

Floating inputs are the opposite of a conflict. Here neither the tester nor the ASIC is driving the net. This may leave the input of the ASIC susceptible to static or other voltage spikes. Additionally, the simulation may not accurately model this somewhat unexpected situation. The simulation may model the floating condition as real data, which are then latched in a register in the ASIC. Clearly, a float should not be relied on to propagate a correct logic value to the ASIC.

Floating inputs are not as severe a problem as conflicts. In fact, floats will occur for a period of time when the tester and the ASIC are changing roles as driver or sensor. However, inputs should not be allowed to float for great lengths of time, and data should never be captured by a sequential device during the time an input is floating.

The first steps of any test program should bring all primary inputs to a known condition and bring all bidirectional pins to a high-impedance state. The second point should be clear from the preceding discussion of conflicts. The reason all primary inputs should be brought to a known state is to avoid the predicament of the simulation applying an X and the tester applying a logic state. Remember that an X represents either a 0 or a 1. Most testers will actually apply a default state when an input is instructed to be at an X. This is usually not a significant problem, although it can cause somewhat unexpected results. For example, the device may in reality receive multiple resets, one during the time the reset pin was an X and one when the designer actually wanted the reset. If measuring the reset time is important, this circumstance invalidates that measurement.

AC Test Methodology

The purpose of AC test is to help verify that the device will run at speed. The measurements made here mimic those that a technician might make at a test bench. The exact types of measurements that ASIC vendors normally provide at this step are highly variable.

> ***Important:* The speed-testing requirements should be discussed with the ASIC vendor to ensure that the normal testing procedures meet the designer's needs. Additional testing is probably possible should it be needed. However, this testing may be available only at an extra cost.**

A common test that is performed at this stage is to measure the propagation time through representative paths in the design. Normally, this means setting up the internal states in the ASIC such that the propagation delays through the longest paths can be measured. Occasionally, this test can be performed along with DC testing if the tester speed and signal integrity are adequate.

It should be noted that AC tests usually require a separate set of test vectors from those used in DC tests: In some cases, it may be possible to simply extract a subset from the DC tests; or the designer may need to generate a completely new set.

DESIGN FOR TESTABILITY

As we stated in Chapter 5 on logic simulation, testability is the efficiency with which a comprehensive test program can be created and executed. Testability, like quality, cannot simply be added at the end of the design process. It must be consciously designed in from the start.

Design-for-testability techniques all have their basis in three simple concepts, initialization, visibility, and controllability, as discussed in Chapter 5.

Ad Hoc Testability Concepts

A number of techniques have been developed over the years to aid designers in improving the initialization, controllability, and visibility of their designs. While a comprehensive review of this subject is beyond the scope of this text, there are a few commonly used techniques that will considerably improve the testability of a design.

Feedback Loops

Feedback loops create considerable problems in test programs and cause longer-than-necessary test sequences. The archetypical feedback loop is represented by a flip-flop in the divide-by-2 configuration. *This circuit will not initialize without direct access to set or reset.* Designers usually respond, "But it doesn't matter." However, it does matter because it will be impossible (or extremely difficult) to synchronize the tester to the ASIC. This is one of the most common design-for-testability mistakes. Figure 8-3 demonstrates the need for a dedicated test reset for this circumstance.

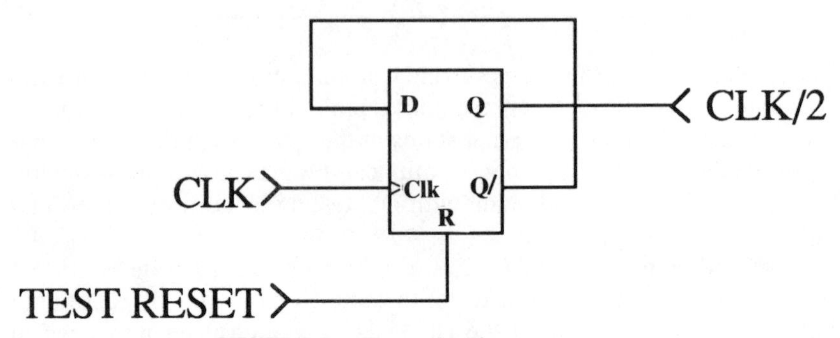

FIGURE 8-3. Example of a test reset.

Counter and Shift Registers

Long count or shift strings create lengthy simulation times and long test times. For example, testing all possible combinations of a 16-bit counter would require more than 65,000 clock cycles. Consider providing the ability to control sections of the string independently. It may be necessary to dedicate pins of the ASIC to test functions that enhance the controllability or visibility of the design. It will be necessary to consider the trade-off of adding dedicated test pins against a longer test development cost if the pins are not used. Proper analysis of such trade-offs is critical to producing an ASIC on time with high quality.

Figure 8-4 demonstrates this concept. TEST_ENABLE allows counter 2 to be counted independently of counter 1. One of the primary

> ***Important:* TEST_ENABLE may be extremely useful to test engineering in the creation of board-level tests. It should be tied to ground on the PCB through a pull-down resistor or, better yet, an inverter whose input is tied to V_{CC} through a pull-up. The inverter is preferred because it is less susceptible to noise than a pull-down.**

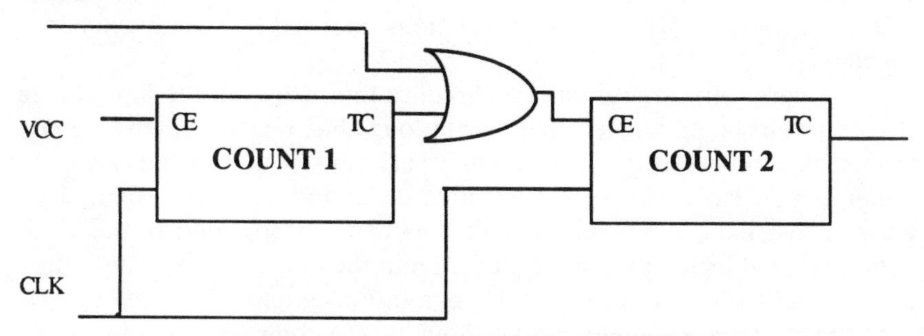

FIGURE 8-4. Breaking long counter chains.

pins of the ASIC must be devoted to TEST_ENABLE. It is important that this pin be tied to ground on the circuit board and not allowed to float.

Asynchronous Events

Asynchronous events, such as using gates to create a pulse (Figure 8-5), create enormous testability problems and should be avoided. These events can make accurately placing tester strobes extremely difficult, since it may be very difficult to determine the maximum propagation delay through the device. Additionally, process, voltage, and temperature variations cause the length of the pulse to vary from device to device, which could decrease the reliability of the test.

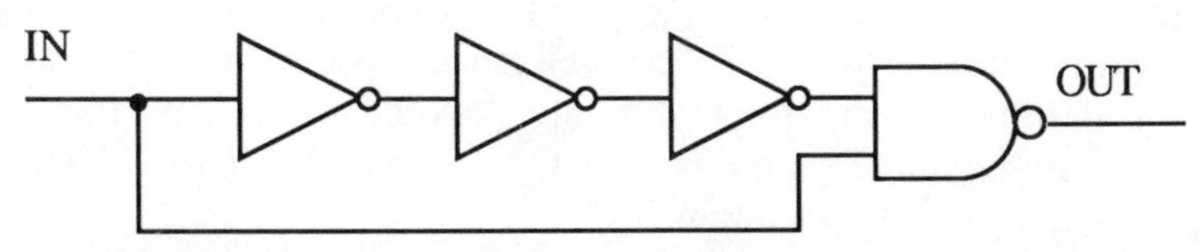

FIGURE 8-5. Avoid using asynchronous pulse generators.

Scan

Design-for-testability concerns and the move toward synchronous design have led to the adoption of more structured design-for-testability techniques, such as scan. Scan design is a methodology that provides high test quality, automation of the test design process, and reduced test turnaround time. To utilize scan fully, however, the designer must adhere to a strict set of design-for-testability guidelines. As ASICs grow in size and as test becomes a significant part of the design process, many designers have become willing to live with the restrictions imposed by scan in exchange for its many benefits.

The basis of scan design is techniques that cause all the flip-flops in the design to be connected together in long shift register chains when a scan enable pin is asserted. These chains are constructed so that combinational logic is isolated between an input chain and an output chain. This permits test data to be shifted into the ASIC, propagated through the combinational logic, and finally latched into the output chain, where they are then shifted out to the tester. From a test programming point of view, a design in the scan mode can be thought of as only combinational circuitry, greatly simplifying the test generation problem. After tests have been generated for the combinational components of the design, the number of shift clocks for inputting and extracting test data can be easily added.

Scan can be utilized with any size ASIC, although it is more common for designs with more than 50,000 gates. Some designers feel, however, that scan techniques should be employed for designs with more than 15,000 gates. A design is a candidate for scan if it is generally synchronous, the ASIC vendor supports scan, 15 to 20% of the ASIC's available real estate can be dedicated to scan testability circuitry, and the analysis tools and tester capability to take advantage of scan are available.

The basis of scan design is scan cells that, when placed in a test mode, are configured into serial chains. A scan cell is both fully controllable and observable. Figure 8-6 shows two common scan cells.

The scan path scan cell is the simplest scan design and requires the least overhead. However, it is inherently race sensitive, and careful balancing of the power buses on the ASIC is required to ensure proper critical path timing. The LSSD scan cell is race insensitive but requires a longer test time and greater real estate devoted to scan on the ASIC. However, the resulting scan design will be race free. Figure 8-7 shows a hypothetical scan design system.

The design is captured as before, except that scan cells are used for the latch components. A scan design rules checker should be used to verify that the design meets the basic criteria for scan. For example, the designer must be careful to avoid using global feedback in a design if scan is to be used. Additional rules may be necessary depending on the robustness of the downstream analysis tools.

***Important:* The fact that an ASIC vendor supplies scan cells in their library is no guarantee that their internal test capability or that of test engineering can utilize it.**

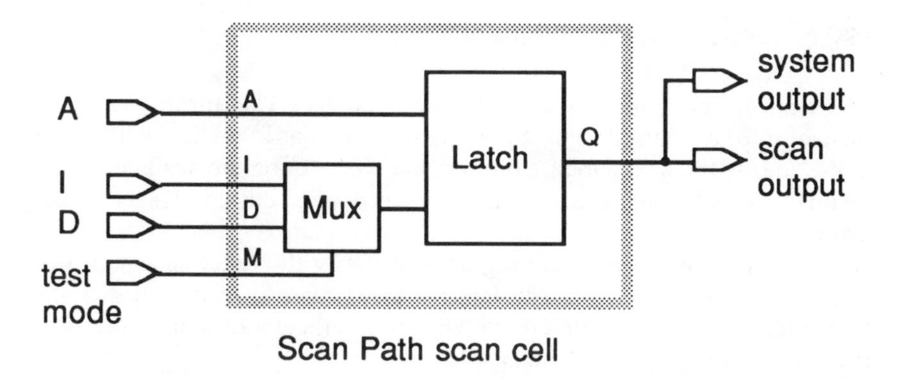

FIGURE 8-6. Scan cells.

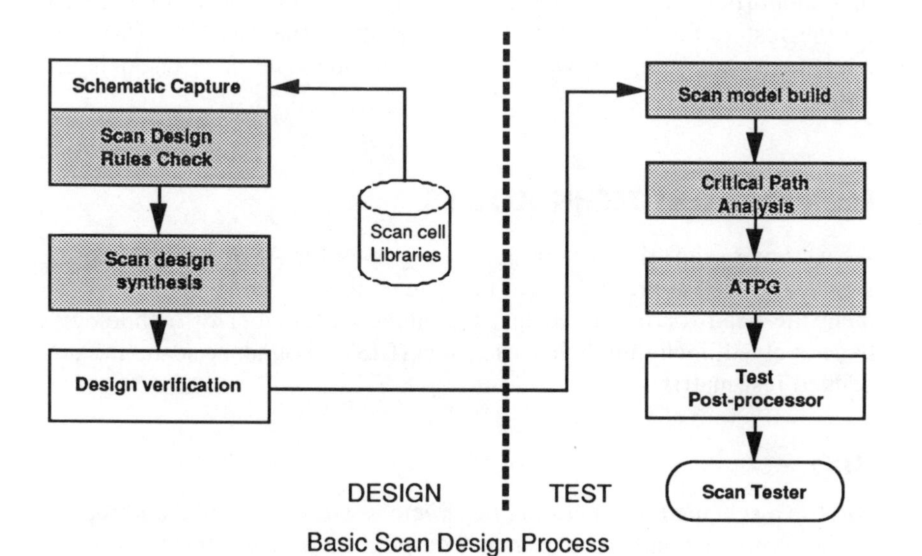

FIGURE 8-7. Hypothetical scan system.

Scan design synthesis may also be available. This turns a nonscan design into a scan design by replacing standard latches with scan cells, adding the scan clock circuitry, and connecting the scan cells into chains. It may be necessary to tune the synthesized design to meet timing criteria.

Design verification is still required to verify the functionality of the design. However, since scan designs are synchronous, zero delay simulation can be employed quite effectively. More efficient design verification is another added benefit of scan.

After successful design verification, the designer is ready to enter the test development phase. Often the design data will need to be optimized for use by the scan tools. After this new scan model is built, critical path analysis is normally utilized to provide timing verification.

An automatic test pattern generator (ATPG) program can then be run to create the scan patterns. This is the point where the benefits of scan are most readily apparent, since patterns can be created in hours that would require weeks or months to generate by hand. The fault coverage created by the ATPG program will be quite high, generally in excess of 99%. The scan patterns are normally then run through a test postprocessor to modify them for the specific syntax of the target scan tester.

Many designers feel that scan is an absolute imperative given the testing demands of today's ASIC designs, and the dramatically increasing popularity of scan is testimony to its effectiveness. However, should you decide to use scan, be certain that you adopt a total methodology that supports scan from schematic capture to testing the finished device on the tester. Failure to do so could result in the real estate used to implement scan going to waste because an adequate scan tester is unobtainable.

EMERGING TEST TECHNOLOGIES

As we have seen, the cost of generating tests for ASIC designs can be significant. However, the cost of inadequate testing can be disastrous. To meet the need for improved test techniques, a number of technologies have evolved, including built-in self-test (BIST), boundary scan, and embedded test matrix.

BIST

BIST is a scheme for generating test vectors within the ASIC and reporting a go/no-go result to a primary output. In general, the term BIST encompasses a wide variety of test methods that meet these criteria. However, BIST is most often implemented through the use of a pseudo-random sequence generator that supplies a repeatable sequence of random data to a block of logic. Results from each test are combined with the result of the previous test through a technique known as signature analysis. After a known quantity of tests have been run against the logic block, a control signal is used to compare the computed signature with a signature programmed into the ASIC by the designer. A matching signature indicates that the logic block passed the test.

BIST is often used to test embedded RAM structures, since a great number of vectors are needed and these are difficult to create using any other technique. In addition, a random testing approach has been shown to be very effective with embedded RAM.

BIST should be designed with great care if it is to be effective. In general, BIST is most useful on extremely regular structures such as memories or purely combinational blocks. In any case, be very careful that only valid data are captured by the signature analysis registers since on the production line a single bit error will be sufficient to indict the device as defective.

Boundary Scan (JTAG)

Boundary scan is a specialized scan structure principally designed to aid with board-level test. Utilizing boundary scan implies the addition of logic to the design to allow the primary pins of the design to be electrically isolated. Data are input and output from the logic normally connected to these primary pins by shift registers, which are enabled by a dedicated test pin on the ASIC. In this way the ASIC is effectively isolated from the rest of the components on the circuit board, allowing test data to be applied to the ASIC without the burden of propagating test vectors through many levels of logic. An added benefit is the ability to identify a defective component on the circuit board immediately, without time-consuming guided fault analysis. In general, boundary scan can be a very effective technique for improving the manufacturability of complex printed circuit boards. Figures 8-8 and 8-9 demonstrate a scan path architecture.

Boundary scan has been highly promoted by an international consortium of companies known as the Joint Test Advisory Group (JTAG). Philips, Texas Instruments, British Telecom, AT&T, and IBM are among the major contributors to JTAG. JTAG has been very active in promoting

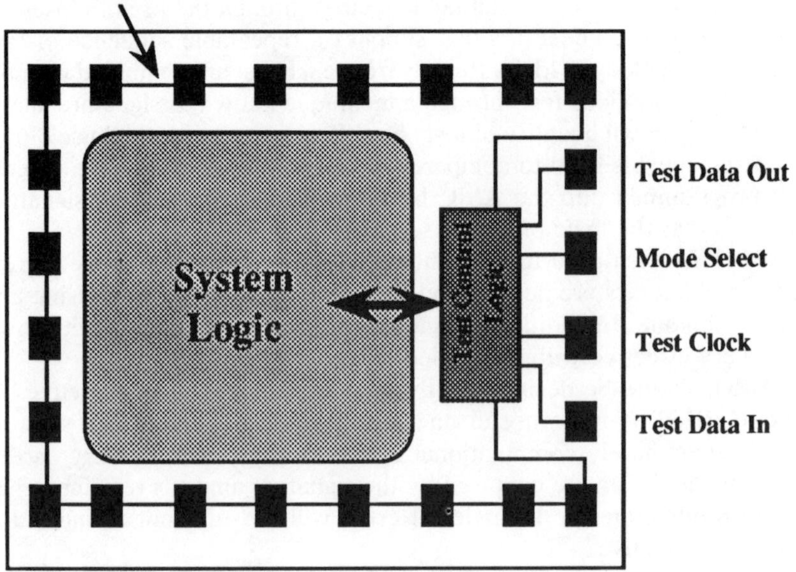

FIGURE 8-8. Boundary scan architecture.

standards in the test industry, particularly in the area of boundary scan. IEEE P1149.1 is a proposed standard for implementing a boundary scan architecture. P1149.1 describes a methodology for implementing boundary scan that minimizes real estate usage and performance degradation. The architecture consists of an instruction register to control switching between test and normal operating modes, a test access port (TAP) for applying test data, and a boundary scan register (BSR).

P1149.1 boundary scan is becoming very well known and popular in the industry as companies seek innovative ways to control test and manufacturing costs. The addition of boundary scan to an ASIC design will cost very little in real estate and can have immense benefits in manufacturing. However, the methodology *must* be consistently applied to the circuit board design to be of any use. Be certain to meet with the circuit board designer and test engineering to verify that the boundary scan design is completely carried through to the board design. It is most disheartening to see a well-implemented boundary scan ASIC mounted on a circuit board with the test circuitry hard-wired to power or ground. Additionally, be certain that the test circuitry cannot be accidentally enabled during normal operation of the circuit board *unless* the board has been specifically designed for self-diagnosis.

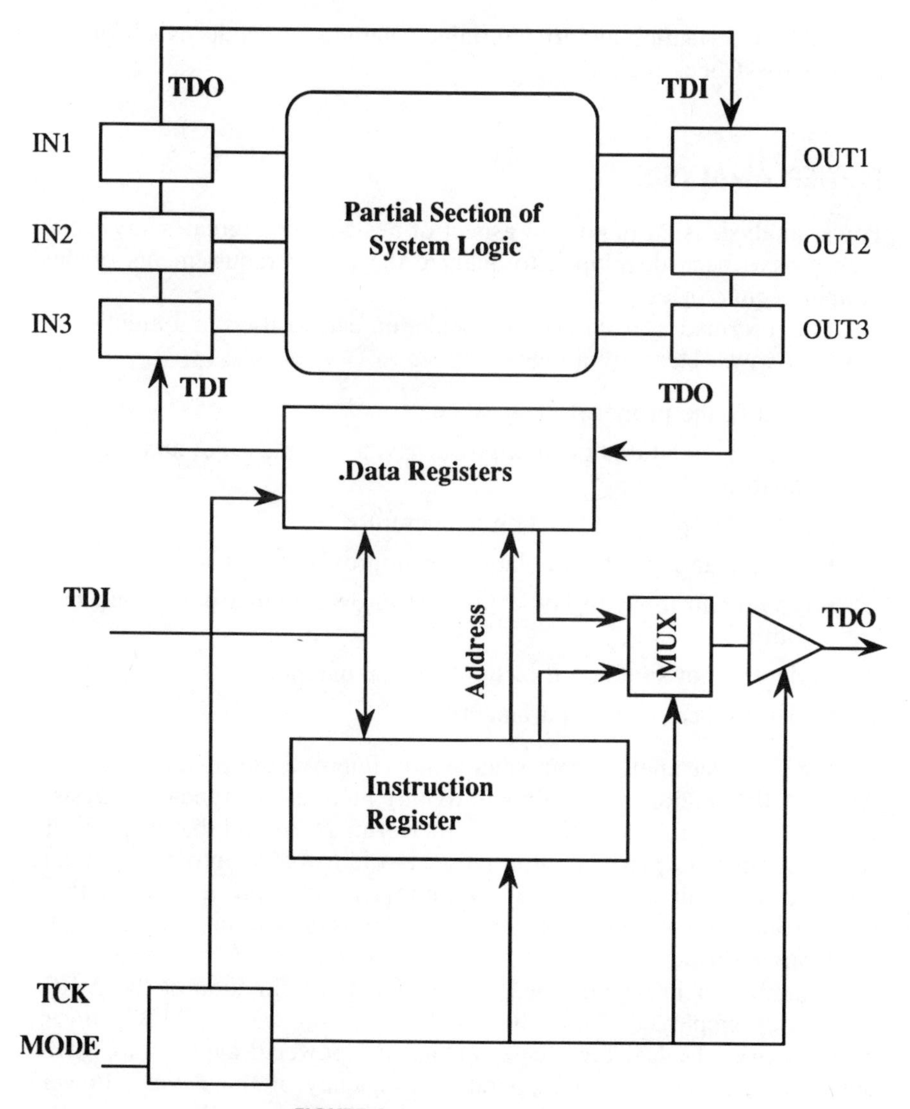

FIGURE 8-9. JTAG architecture.

Embedded Test Matrix

One of the most promising of the emerging test technologies is embedded test matrix, which is being developed by the CrossCheck corporation. This proprietary technology permits every node in the design to be completely observable, allowing for extremely high fault coverage with normal functional test techniques. Check with the ASIC vendor for up-to-date information and to determine whether the vendor is a licensee of the CrossCheck technology.

POWER ANALYSIS

Power analysis is an important aspect of design, and a great many techniques have been developed to analyze the power requirements of the various technologies.

The information from power calculation can be used in a number of different ways. Here are a few:

- Identify the proper package type.
- Identify violations of total power given die, package, and thermal constraints.
- Estimate operating junction temperature.
- Provide an input to simulation for improved simulation accuracy.
- Provide an input to layout for use in determining power and bus widths.
- Identify hot spots on the chip (thermal mapping).
- Identify heat sink requirements.

Since power analysis can significantly improve the reliability of the ASIC, ASIC vendors are tending to require more of this type of analysis. In many cases, the calculation will be performed by a tool that the vendor supplies. Specific process information is often folded into the general power calculation to increase its accuracy. In addition, it is likely that tools from commercial workstation vendors will become increasingly available in this area.

The manner in which power is calculated is a function of the ASIC technology employed. There has been much interest in CMOS power analysis over the last few years, in that the power drawn by a CMOS device is a function of the operating frequency of the device. Power consumption in some technologies, such as bipolar and gallium arsenide, is basically insensitive to operating frequency. Let us look at these two types of power analysis.

CMOS Power Analysis

CMOS power consumption is a function of voltage, frequency, capacitance, and load. Total power is the sum of the static power drawn from the power supply and the switching power of each device in the design. Average dynamic power of a CMOS cell can be calculated by the equation

$$P_d = (C_l + C_{pd})(V_{cc2})f$$

where:

f is the average frequency in megahertz.

C_l is the total load capacitance on a cell output in picofarads. C_l is a function of net capacitance and input capacitance.

C_{pd}, the power dissipation capacitance, is the internal device capacitance in picofarads.

The value of f can be calculated from the equation:

$$f = [(\text{Toggle\_number})/2](\text{Total\_time})$$

The toggle number is the total number of output pin transitions. Total time represents the total amount of simulation time. Therefore, f can be computed by simply counting the number of times an output transitions and dividing by the total simulation time. C_{pd} is a characteristic of each design cell and must be provided by the ASIC vendor. Since C_l is a function of layout, its value must also be provided by the ASIC vendor. C_l will normally be calculated as part of the process of acquiring layout-specific timing information. See Chapter 6 for more details on the annotation process.

Bipolar and Gallium Arsenide Power Analysis

As mentioned, power consumption in bipolar and gallium arsenide (GaAs) devices tends to be frequency-independent. Therefore, total power consumption is a function of the number of and type of design cells. The power consumption of an individual cell can be determined from the data sheet supplied by the ASIC vendor.

$$\text{Total power} = (\Sigma \text{ Number of instances})(\Sigma \text{ Power of each instance})$$

For example, if the design consisted of 1,000 gates that consumed 4.5 mW and 2,000 gates that consumed 2.2 mW, the total power would be

$$\text{Total power} = (4.5 \text{ mW})(1000) + (2.2 \text{ mW})(2000) = 8.9 \text{ W}$$

BOARD TEST

The test data needed to verify the integrity of the prototype ASICs can probably also be used by test engineering for development of board-level tests. By leveraging this information, test engineering can deliver a high-quality test in less time, helping to speed the product to the manufacturing floor.

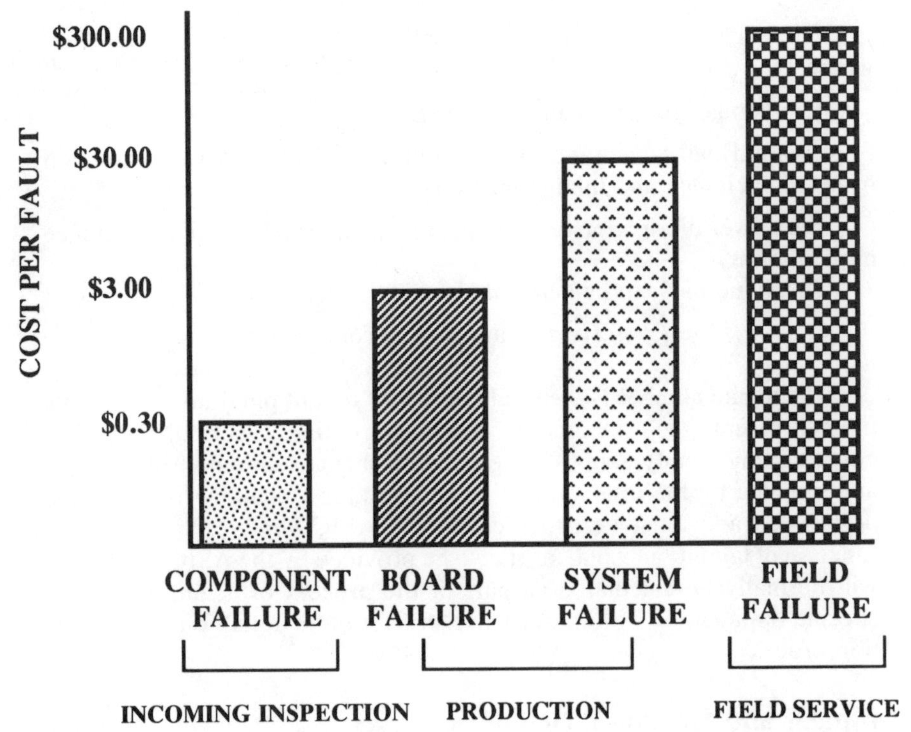

FIGURE 8-10. Test cost at each step in the manufacturing process. (© Prime Data.)

Figure 8-10 represents the average cost of finding and repairing failed components at each step in the manufacturing process. Notice that finding a defective ASIC in the field is one thousand times more expensive than finding the same defective ASIC at incoming inspection. It is easy to see the benefits of detecting defective devices as early in the manufacturing process as possible.

The testing philosophies of companies vary significantly. If you have never taken a trip to visit the test engineering group, this is probably a good time to do so. Remember, this organization can significantly affect the success of a design by developing tests that detect defects early on.

One question for the test engineers is, Do you do an incoming inspection test? Some companies believe that IC vendors are responsible for ensuring the quality of the devices they deliver. Therefore, incoming inspection is not performed. These testers are quite expensive ($1 million is not unusual), so this philosophy can save the company considerable expense. Some companies conduct incoming inspection on only a selected sample of devices with the goal of recognizing quality trends. There are independent test services that provide this service as well. There is no right or wrong here. Whether or not to conduct incoming inspection is a business decision based on the price of equipment and personnel versus the cost of detecting defects later in the manufacturing process.

It is quite likely that the company owns some type of automated board tester to test the completed printed circuit board assembly. Ask the test engineers whether the board the ASIC plugs into will be tested on one of these pieces of equipment. If so, the designer has the opportunity to greatly assist test engineering in producing the highest test possible and reducing the chance of a defect reaching the field.

There are two general classes of automated board tester, in-circuit and functional. Combination testers, which have both in-circuit and functional capabilities, are also available. We will define in-circuit and functional test briefly and describe the benefits of each so that you will have a base level of knowledge as you discuss board test needs with test engineering.

In-circuit Test

In-circuit testers make a connection to every node on the circuit board through a vacuum test fixture commonly called a bed of nails. The goal of an in-circuit test is to detect common manufacturing defects as early in the manufacturing process as possible. In-circuit testers are very adept at detecting shorts, opens, missing components, and dead components. They are less likely to find subtle functional failures or timing defects.

Each device on the board is tested individually by forcing the inputs to the devices to logic values specified in a test program for that particular type of device. High-current drivers are used to force the nodes to the desired values. Numerous studies have shown that this technique will not harm devices for the short period the test is run.

The test program sent to the ASIC vendor for prototype verification can probably serve as the basis for the in-circuit test of the device. It may be necessary to shorten the test sequence since these testers generally have less pin memory than a device tester.

In-circuit testers have become popular for a number of important reasons:

- They are easy to program.
- Their diagnostics are very good.
- They are less expensive than functional testers.
- They detect common manufacturing defects efficiently.
- They do not require highly skilled operators.

Functional Test

Functional testers provide stimulus to the primary inputs and make measurements at the primary outputs of the printed circuit board. The goal of a functional test is to test the board in an environment that matches the backplane the board plugs into as accurately as possible. Logic simulation is often used to aid in creation of the expected output responses.

Creating stimulus for a functional test program is a very difficult process that depends on an intimate knowledge of the system and the board. The designer's expertise is of great interest to the test engineer creating a functional test program. The simulation model of the ASIC may prove quite useful, particularly if the simulators used in test engineering are the same as those used in design engineering. This is one reason companies are moving to standardize logic simulators throughout the organization.

The benefits of functional test are:

- Timing errors and subtle defects can be detected. These defects can be quite difficult and time consuming to detect at system test.
- Fixturing requirements are much simpler than those for in-circuit test since connections need be made only to the primary pins of the circuit board.
- Box-level tests are generally avoided. The next level of test is normally system-level.

Diagnostics in a functional tester are normally accomplished with a technique known as guided probe. The tester will ask an operator to physically probe the device inputs that affect a failing output. If one of these input nodes is determined to be bad, the tester then asks the operator to probe the nodes that affect this node. If all inputs that affect a failing output are found to be good, the tester will indict the component producing the failing output as defective.

SUMMARY

The development of a test plan early in the development process is critical to ensuring that the ASIC meets the levels of quality required. Discuss the requirements with the ASIC vendor and the in-house test engineering and quality organizations. Use design-for-testability techniques, and consider scan design where appropriate.

REFERENCES

Abramovici, Miron, Melvin Breurer, and Arthur D. Friedman. 1990. *Digital Systems Testing and Testable Design* pp. 343–408, New York: Computer Science Press.

Archambeau, Eric, and Ken Van Egmond. 1988. Built-In Test Compiler in an ASIC Environment. In *1988 International Test Conference Proceedings* pp. 657–663, New York: IEEE.

Beausang, James, and Alexander Albick. 1985. Incorporation of the BILBO Technology Within and Existing Chip Design. In *IEEE 1985 Customer Integrated Circuits Conference* pp. 326–332. New York: IEEE Publishing Services.

Cron, Adam D. The IEEE-1149.1 ASIC—Going Further Than Boundary Scan. In *1990 Test Engineering Conference Proceedings* pp. 37–44. Boston: Miller Freeman Expositions.

Culbertson, Gary, 1988. Managing the ASIC Design to Test Process. In *1988 International Test Conference Proceedings* pp. 649–656, New York: IEEE.

Davis, Brendan. 1982. *The Economics of Automatic Testing*. London: McGraw-Hill (UK) Limited.

Fujiwara, Hideo. 1985. *Logic Testing and Design for Reliability* pp. 133–144, 206–267. Cambridge, MA: MIT Press.

GigaBit Logic, Inc. 1989. *SC10000 Standard Cell Array Design Manual*. Newbury Park, CA: GigaBit Logic, Inc.

LSI Logic Corporation. 1988. *LSI Logic Primer for ASIC Design*. Milpitas, CA: LSI Logic Corporation.

LSI Logic Corporation. 1989. *LSI Logic Chip-Level Full-Scan Design Methodology Guide*. Milpitas, CA: LSI Logic Corporation.

Maunder, Colin M., and Rodham E. Tulloss. 1990. The Test Access Port and Boundary-Scan Architecture. In *1990 International Test Conference Proceedings,* pps. 44–51, Los Alamitos, CA: IEEE Computer Society Press.

Parker, Kenneth P. 1987. *Integrating Design and Test: Using CAE Tools for ATE Programming,* Washington: Computer Society Press of the IEEE.

Remeis, Paul. 1990. Applying Fast Fault Grading Techniques Benefits Design and Test Engineering. In *1990 Test Engineering Conference Proceedings* pp. 105–118. Boston: Miller Freeman Expositions.

Turino, Jon. 1990. Testability Impact on Reliability. In *1990 Test Engineering Conference Proceedings* pp. 1–10. Boston: Miller Freeman Exposition.

Turino, Jon. 1990. *Design to Test*. New York: Van Nostrand Reinhold.

Waicukauski, John, Paul Shupe, David Giramma, and Arshad Matin. 1990. *ATPG for Ultra-Large Structured Design,* Alamitos, CA: IEEE Computer Society Press.

9

Summary and Conclusions

INTRODUCTION

The most important point that this book makes is that the design of an ASIC is not a simple task. It is, however, an essential part of today's electronics development process, so that at least a fundamental knowledge of the ASIC design process is a priority for every designer. The ASIC itself can be a complex device and the design tasks in total can be complex, but the techniques explained in this book have shown the new ASIC designer the right approach to use to get the design completed correctly with a minimum of complications.

ASICs can be made up of all digital cells, all analog cells, or a combination of analog and digital cells. Special processing techniques must be used for analog and mixed analog-digital ASICs to ensure that accuracy and performance are maintained, and only a few ASIC vendors have the experience and processing techniques required to bring these complex devices to completion.

This book contains information about the process of selecting an ASIC vendor that will meet your needs, both near-term and long-term.

WHAT YOU HAVE (OR SHOULD HAVE) LEARNED—A REVIEW

The ASIC design process varies depending on the ASIC vendor used *and* the internal development guidelines defined by the designer's company. Of all the ASIC design processes that were reviewed, no two were identical. However, from these processes a general flow could be derived. (See Figure 2-1.) Each ASIC vendor has a specific design process that was developed to support their process tools. This process has been refined over the ASIC vendor's lifetime by developing ASICs. As geometries and technologies changed, so did the development process, and the tools that the designer uses were adjusted to meet new requirements.

The ASIC process will depend on the end application of the ASIC being designed. If the application is a mainframe computer, then high-performance characteristics will be of primary concern. If the application is a personal computer, then high-density technology combined with high-performance technology will be key. The end application will also play a large role in determining the right ASIC vendor, the technology to use, and the packaging scheme required.

Regardless of the end application, the design must be completely specified before any design begins. Design specifications must include the target board environment as well as the details of the ASIC. The specifications must have estimates of layout effects on timing parameters to help with design characterization during simulation. Today's design tools allow complex ASICs to be specified and simulated at a behavioral level using VHDL. After the design is verified at the behavioral level using VHDL, synthesis tools can be used to transpose the design into structured logic gates for detailed design verification.

Packaging of the ASIC should be determined early in the development cycle. Too often, the packaging is determined late, causing delays and in some cases redesign. Defining packaging requirements at the beginning of the design as a normal part of the design specifications can save time, development costs, and frustration. There are many package types to consider, and a designer cannot afford to change the packaging of an ASIC after the board has been through layout. Typically, the number of pins and the technology used determine the package type needed. Different package types have different mounting area requirements. Therefore, the designer should have a good understanding of the mounting area available on the target board for the ASIC package.

All ASIC designs are simulated before layout. The simulation steps consist of functional and timing simulation to verify the logical operations of the ASIC design. Further, all ASIC designs are fully fault-tested to verify that all test vectors for the design can test all inputs and outputs.

WHEN TO USE THE ASIC VENDOR'S DESIGN TOOLS AND WHEN TO BUY YOUR OWN

One of the benefits derived from using an established ASIC vendor is the flexibility that they can offer a designer. A design can be created and fully simulated using either the designer's own engineering workstations or the ASIC vendor's design center workstations. Conversely, any portion of the front-end process can be performed using a combination of the designer's EDA workstations and the ASIC vendor's EDA equipment.

Most ASIC vendors support to some degree at least the top three EDA vendors' workstation environments, including full library support for both gate arrays and standard cells. However, the designer should verify that the libraries supported are using the process technology required for the ASIC design. Some ASIC libraries are behind on the EDA vendor's latest software releases and may not match the designer's cell needs. ASIC vendors try to maintain their top two or three EDA vendors' libraries, but history has shown that they will always support the industry leader first.

Q. Should more than one ASIC vendor be used?

A. One ASIC vendor can certainly meet the needs of almost all ASIC designers on a per-project basis. However, not all ASIC vendors can support all design needs for every project. Some ASIC vendors are excellent at digital but have no experience at analog ASIC design. Conversely, ASIC vendors who have developed their business with analog ASIC technology will not be able to meet digital design needs. Then, there is the latest group designing mixed analog-digital. There are very few ASIC vendors who can provide the design and process tools needed to develop mixed analog-digital ASICs. This means that the designer may have to use more than one ASIC vendor to get a project successfully completed. Many factors, including track record, design-support staff experience, and foundry facilities, will contribute to identifying whether the ASIC vendor can cover all the design requirements.

How does a designer find the *ideal* ASIC vendor? What does the ideal ASIC vendor look like?

HOW TO CHOOSE AN EDA VENDOR

About ten EDA vendors dominate the design automation market, having over 75% of the total. These EDA vendors offer everything from basic schematic capture to broad-based board/system design environments. Choosing an EDA vendor is very similar to choosing an ASIC vendor. The designer must check track record, tool specifications, and support capabilities to ascertain whether a particular EDA vendor can meet his or her needs. In some cases, depending upon the application, more than one EDA vendor may be required to enable a design project to be completed. In these cases, the designer must evaluate several areas where data interfacing will be required to ensure compatibility. The designer should ensure that data can be moved bidirectionally between the EDA systems involved. Areas of concern are:

- Schematic capture
- Moving completed schematics between systems via netlists or translator tools for simulation or layout
- Back annotating the schematic design with layout-derived timing information to verify that actual design parameters do not cause the design to fail
- Simulation
- Moving test vectors between systems to avoid having to re-create them
- Moving simulation data to a hardware simulation accelerator and the results back
- Models
- VHDL model compatibility among systems
- Synthesis of behavioral language models into compatible structured models for simulation or layout

Choosing the EDA vendor that will provide the tools and support to ensure that projects are completed in a timely manner while maintaining the highest level of quality is not an easy task. The choice will undoubtedly require a full-scale product/tools evaluation involving several EDA vendors and will take several months to complete. Indeed, the design team must be prepared to take on the task of developing benchmark test cases that the EDA vendors must complete to be considered for the business. Proper planning at the beginning will prevent underestimating the time needed to put the benchmark test cases together. The benchmark should be generic enough to allow all EDA vendors an equivalent chance to complete it. Some EDA vendors will not be able to complete the entire benchmark or will provide incorrect results. It is common for the design team to introduce problems into the benchmark that have taken hours to uncover in the normal design environment in order to identify weaknesses in the EDA vendors' tools.

There are three basic types of design environments, and they require different methods for choosing EDA tools. Group 1 is a design team that has little or no experience with EDA tools. Group 2 is a group that has experience only in specific segments of EDA tools, such as schematic capture. Group 3 is a group that has been using EDA tools but needs to find out if a change in environment is required. Regardless of the group, the design team must define the requirements that are needed to ensure the success of their development project. A suggested requirements list should be divided into must-have, should-have, and would-be-nice categories. In addition, there should be a section covering future requirements that are not currently needed.

There are several ways that a design team can make the necessary selections to fulfill their requirements. One method is to publish their requirements documents and ask interested EDA vendors to provide written responses pertaining to how each requirement is covered. Another may be to use the internal requirements document as a basis for a full-scale benchmark evaluation of EDA vendors and measure how each vendor performs against the criteria. Either method is a time-consuming endeavor requiring dedicated engineering resources.

LONG-TERM VIABILITY OF ASIC METHODOLOGY FOR BOARD-SYSTEM DESIGN

The basic premise is that ASICs in some form will continue to be the mainstay technology driver of the electronics design environment. ASIC technology will continue to evolve and challenge the custom integrated circuit for supremacy for the development of integrated circuits. Today's gate arrays are rapidly being replaced by both standard cells and PLD technology. Standard cells are replacing custom IC designs as the number of complex cells available increases and as the use of synthesis techniques provides higher densities. The future looks very promising for the ASIC coupled with design synthesis for layout, as these techniques are driving cell compaction toward that of a custom IC. As these technologies mature, costs will continue to decrease, even for large quantities, causing more efforts to be put into cell-based design.

FUTURE OF EDA TOOLS AND METHODS

The EDA market was born from the integrated circuit technology boom of the 1970s and 1980s. This foundation will continue to drive EDA vendors to invest in research and development for tools that will increase design productivity with integrated circuits. At the same time, EDA vendors will continue to push R & D to improve the design productivity of the integrated circuit's target environment, the circuit board. Finally, EDA will continue to provide tools that will ease the design and integration of the IC, board, and system environments. EDA offers the only way to increase design productivity while making the designer's job easier, across the electronic design spectrum.

Some of the areas that will provide the productivity and integration needed for the design of ICs, boards, and full systems are:

- *Concurrent design* is a method that a large number of electronics companies have been using for many years. EDA vendors have slowly begun to support these methods as a normal part of their design tools environment. Concurrent design support from the EDA industry will provide links to downstream processes, allowing engineers and their managers to see the impact of design decisions early in the design process. The EDA vendors' support of concurrent design will result in tool linkages that allow trade-offs to be explored and changes made where the costs are lower.
- *Open architectures* will provide the means for tightly integrating proprietary tools or tools from a competitive EDA vendor in order to provide the right solution for the designer or design team. Open means allowing both forward and backward design data exchanges to support the concurrent design of ICs, boards, and systems.
- *Frameworks* for plugging in nonproprietary EDA tools will provide the base technology that defines the architectural guidelines for connecting proprietary and competitive EDA solutions into an EDA vendor's environment. A design framework must provide decision support, design management, and an accessible object-oriented database for efficient data sharing.
- *Higher-performance design tools* geared to new ASIC technologies will continue to be developed to support the increasing complexities of ASICs, boards, and systems. These EDA tools will focus on design specifications, design synthesis, simulation, layout, and test. All EDA tools that support these areas must continue to keep pace with ASIC technologies.

BOARD AND SYSTEM SIMULATION

These are the areas that will see the most attention from EDA vendors over the next decade. The areas of board and system simulation are important to ASIC designers because they represent the final target environments for each ASIC being developed. A tighter integration of ASICs and boards for simulation will provide increased performance and reliability and will get the product to market much more quickly.

Areas of concern to both designers and EDA vendors fall into several categories: Will code execution be a reality? Realtime? How fast?

As microprocessor cores continue to advance the state of the art for cell-based design, the question of realtime code execution becomes more important. Designers need the ability to verify several minutes of realtime operations when performing a simulation of a complex ASIC. This task has not been an easy one for EDA vendors to solve. Two approaches that have been used are the use of special-purpose hardware simulation engines and the use of hierarchical, high-level models.

The use of special hardware has provided a boost in simulation runtime performance of two to three orders of magnitude over a generic workstation when running at the structured gate level. These machines are known as hardware simulation engines or gate-level simulation accelerators. These hardware simulation engines can perform both logic and fault simulations on digital designs. Many have been adapted by ASIC vendors to offset mainframe computer workloads. Recent advances in hardware simulation technologies have provided the ability to run multilevel simulations, where a mix of structured and behavioral models can be evaluated simultaneously. The rapid evolution of workstation performance technology has begun to challenge hardware simulation engines, especially in multilevel simulation applications.

Regardless of the choice, today the ability to execute large strings of code in a reasonable time does not exist, nor is it likely to for the next several years. Perhaps by the end of the decade a solution will be found.

The other hardware simulation solution was the hardware modeling system, which provided a simulation interface between the digital workstation simulator and a real integrated circuit. The real integrated circuit was plugged into a special circuit board where it could be stimulated with inputs from the digital simulator and then provide a response to the simulator in the form of time-synchronized events. This technology allows an ASIC (prototype or production) to be used during a full board simulation to verify functional simulation of an entire design. Hardware modeling systems use a designer-defined timing information lookup file that the simulator accesses based on the response needed. Since these files are generated by the designer, hardware modeling offers a means to stress the entire design based on the timing parameters assigned to the ASIC. By simply changing parameters in the timing file, the ASIC will perform accordingly during a simulation run.

THE FUTURE OF MIXED ANALOG-DIGITAL ASICS

The mixed analog-digital, or mixed-signal, ASIC is continuing to be used more frequently by designers. Industry sources have shown that more than two-thirds of all circuit boards produced today contain both analog and digital circuitry.

Still, estimates show that over 75% of mixed-signal ASICs have 80–90% digital circuitry. According to industry sources, designers will be using more mixed-signal circuitry in order to provide more capabilities in less space using less power.

The biggest controversy revolves around having all the right tools to allow designers to correctly develop, model, and verify mixed-signal ASICs. What will the EDA vendors offer designers? What are ASIC vendors doing to handle this new outbreak of technology? What must the "new" ASIC designer understand about mixed-signal designs?

EDA Vendors

The focus needed by EDA vendors is broadly segmented into three categories: analog VHDL, analog synthesis, and mixed-signal simulation.

These three areas are in the early development stage at most, but are receiving R & D funding by major EDA vendors. Product plans, panels, and industry committees have been developed to guide the efforts for mixed-signal tools.

- *Analog synthesis* is the ability to take high-level behavioral descriptions of analog models and transform them into transistor- or gate-level representations for either pure analog or mixed-signal simulation.
- *Better mixed-signal simulation tools* are needed to ensure the accuracy and reliability of mixed-signal ASIC designs. This means that patching digital and analog simulators together will not be a viable, cost-effective solution for real mixed-signal ASIC designs. The requirement is to provide mixed-signal ASIC designers with a truly integrated mixed-signal simulation environment.
- *Analog HDL* is the next frontier for EDA vendors. Providing a high-level language-based standard for analog models will enable designers to develop complete, accurate, and reliable mixed-signal ASICs more quickly.

ASIC Vendors

The challenge for ASIC designers is going to be to try to keep up with the demand for mixed-component technology. Designers will require a selection of cells comparable to digital cell offerings. Most ASIC vendors do not even carry analog cells yet. There is now a push to add digital signal processing (DSP) core cells to better assist ASIC designers performing complex audio and video frequency applications.

The real goal of ASIC vendors should be to push designers into using off-the-shelf analog cells. The solution is to provide a sufficient offering of analog cells, both standard and customizable, to meet mixed-signal ASIC design needs.

Vendors will have to help to enhance the process technology used for mixed analog-digital, especially CMOS, as it currently puts some severe limitations on the resolution and accuracy of analog cells. This shows up primarily in high-speed analog functions. At the same time, high-load simultaneous switching transients can cause the ground potential to change in analog cells.

Analog characteristics, and therefore the cell yield, can be affected by the process inconsistencies of interconnects and base materials. In addition, an area that needs close attention is the fabrication techniques for capacitors. Current techniques do not provide the reliability and accuracy required for today's or tomorrow's mixed-signal ASIC designs.

ASIC Designers

Designers must continue to push both ASIC and EDA vendors to supply the technology, processes, and tools to allow ASIC designs to expand. One area where designers need to have more flexibility is with analog cells. They need to be able to modify existing cells or to have the ASIC vendor create new cells (macro functions).

Designers are struggling with technology limitations such as those with CMOS that affect noise margins, gain characteristics, and voltage referencing. Designers must put pressure on the ASIC vendors to help with overcoming design-limiting technologies. Additionally, designers are finding themselves faced with limited cell libraries from some ASIC vendors, forcing them to find and use vendors who specialize in analog cell design and processing. This process causes severe complications with mixed analog-digital ASIC design and limits the design to an expensive hybrid IC approach.

REFERENCES

Goodenough, F., Design Analog ICs with CAE/CAD System. *Electronic Design,* 1989.

Pryce, D. Analog-digital ICs Provide Versatility. *EDN,* 1990.

Glossary

Courtesy of Mentor Graphics Corporation

Analysis Timestep: *See* Timestep.

ASCII: The American Standard Code for Information Interchange. A standard character set used throughout the computer industry.

ATE: Automated Test Equipment. A system that applies test patterns to a piece of hardware to verify its functionality.

Back End: A generic name for a simulator's kernel or simulation server.

Behavioral-level Model: A simulation model whose behavior is specified algorithmically. BLMs are behavioral-level models.

Bi-directional: A characteristic of a pin or model that allows it to pass signals in either direction.

BLM: Behavioral Language Model—a C or Pascal object file that models the behavior of a part. A BLM describes the part's function at the algorithmic level.

Breakpoint: A means by which you can interrupt the progress of a simulation, based on specific behavior of the simulation. For example, you could set a breakpoint to interrupt the simulation when a particular signal goes high and another signal goes low.

Bug: An unwanted and unintended property of a program. Something that causes it to operate incorrectly. (From telephone terminology, "bugs in a telephone cable" blamed for noisy lines; also from the moth that flew into ILLIAC I and shorted out all the circuits.)

Bus: A group of related signal lines combined on a schematic or in the simulator.

C: A high-level programming language, developed by AT&T Bell Laboratories, which performs certain types of binary operations that previously could only be done in assembly language.

CAD: *See* computer-aided design.

CAE: *See* computer-aided engineering.

Computer-aided Design (CAD): The use of computer-based tools to assist in the physical layout of electronic designs, including preparation of manufacturing tooling. CAD also refers to automated mechanical design.

Computer-aided Engineering (CAE): The use of computer-based tools to assist in the creation of electronic designs, from initial specification through layout, analysis, and production.

Concurrent Fault Simulation: A type of fault simulation that simulates multiple faults while simultaneously performing a fault-free (good-circuit) simulation. There are no restrictions on modeling methods, and the simulation speed is faster than with serial or parallel fault simulation.

Controllability: The ability to manipulate the state of a signal within a circuit by applying stimulus to the primary inputs of the circuit. *See also* Observability.

Coverage: The extent to which a set of test vectors can detect faults in a circuit. Coverage is expressed as a percentage of all conceivable faults in the circuit.

Cycle: In fault simulation, the duration of time during which the simulator applies input stimulus, reads the output response, and waits for the next application of input stimulus.

Decay: In logic simulation, the condition that occurs when the voltage charge (trapped at a pin) leaks away, resulting in a subsequent change in the pin's logic level.

Delay: *See* Inertial Delay, Propagation Delay, or Transport Delay.

Design Database: A set of files that describe a design.

Design Hierarchy: A means of representing increasing levels of detail in a design. For example, the top level of a design hierarchy could be a block diagram, the next level could be the logic gates that implement the block diagram, and the bottom level could be the transistors that implement the logic gates.

Design Validation Tests: A set of input stimuli applied to a circuit to test for proper operation.

Device: A physically distinct hardware component, such as an IC or transistor.

Distribution List: A list of nodes that participate in the current LAN simulation. *See also* LAN Simulation.

Drive Strength: In digital logic simulation, an indication of the rate at which a signal source can provide or delete a voltage charge on a net.

EDA: Electronic design automation.

Evaluation: The act of a simulator determining the output of an instance based on the instance's input values.

Event: In simulation, the change in a state value at an output pin of an instance.

Fault: A manufacturing defect that causes an electronic circuit to perform incorrectly. Examples of faults include a solder bridge (short circuit) and a broken wire (open circuit). A fault simulator or grader models faults in a simulated design.

Fault Collapsing: The process by which a simulator combines equivalent faults prior to fault simulation.

Fault Coverage: *See* Coverage.

Fault Detection: The process by which a fault simulator discovers the existence of faults. The simulator detects a fault when input stimulus causes a primary output to differ from the good-circuit response.

Fault Grading: The use of computer tools to simulate the behavior of a design and use statistics gathered during the simulation to predict the probability of faults occurring in that design. The purpose of fault grading is to determine how well a test program can detect manufacturing defects.

Fault Insertion: The process by which a fault simulator places faults in the simulated circuit.

Fault Simulation: The use of computer tools to imitate the behavior of designs that contain manufacturing defects. The purpose of fault simulation is to determine how well a test program can detect manufacturing defects.

Functional Tests: Sequences of input stimuli applied to the evolving design, in order to verify that the design functions according to specification.

Gate-level Model: A simulation model that is composed of primitive logic gates.

Hardware Modeling: A technique that allows the designer to use real IC devices as simulation primitives, thus avoiding the need to write a software description of a complex device, such as a microprocessor.

Hazard: A simulation error that occurs with zero-delay gates when the simulator schedules two conflicting signal values at the same timestep on the same output pin.

High Activity: Those faults that were often controlled to logic zero and logic one and had a high probability of being observed at a primary output. They are faults that a fault simulator would likely detect.

High-impedance Strength: One of the possible signal strengths that a simulator can model. When a signal has high-impedance strength, that signal can neither add nor remove a charge on a net. *See also* Drive Strength.

IEEE: Institute of Electrical and Electronics Engineers. This incorporated organization creates and manages industry-wide standards such as the IEEE Standard VHDL Reference Manual designated as IEEE Std 1076–1987. *See* VHDL.

Inertial Delay: A type of propagation delay where the instance does not accept input signals that transition faster than the instance's propagation delay; instead, the instance generates a spike warning. *See also* Spike, Propagation Delay, and Transport Delay.

Initialize: The act of setting logic elements in a digital design to a known state.

Interactive Simulator: A simulator that allows you to halt simulation at any point in order to investigate the state of signals in the circuit or change input stimulus, and then continue simulation from that point.

JEDEC: Joint electron device engineering council. This council sets standards for electronic manufacturing such as the JEDEC standard that defines how a PLD fuse map (programming file) should be formatted.

Kernel: *See* Back End.

"Known Good" Component: The implementation of a circuit that is defined as being correct.

LAN: Local area network. A means of connecting together a group of computers so that they can communicate to each other, as in an Apollo network.

LAN Simulation: A concurrent fault simulation that is distributed over multiple nodes in a LAN (local area network).

Large-Scale Integration: Standard digital integrated circuits (chips) having approximately 1000 transistors.

Layers: Logical CAE/CAD/CAM data divisions that can be viewed individually or as overlays.

Layout: That step in the electronic design process during which the definition of the circuit is transformed into a physical equivalent. This equivalent can be manufactured as a gate array, a standard cell, a custom IC, or a PCB.

LCCC: Leadless ceramic chip carrier. A hermetically sealable ceramic package in which an integrated circuit chip can be mounted to form a surface mount component. Instead of leads it has pads around its perimeter for connection to the substrate.

Legal Location: A location for a device or component that is deemed acceptable. This can be used as a reference point.

Library: A collection of often-used electronic parts that typically contain the electrical characteristics used in the different phases of electronic design.

Logfile: An ASCII file that contains a record of circuit activity that occurred during a simulation run.

Logic Design: To create an electronic design using predefined building blocks such as gates (AND, OR, NAND, . . .).

Logic Diagram: A graphic representation of the logic devices in a computer, such as flip-flops, gates, and circuits.

Logic Element: A symbol that has logical meaning; also called a logic symbol. For example, the symbol for an inverter (a triangle with a small circle on the output end) represents a device that inverts the input signal (changes from a one to a zero or from a zero to a one) and transfers it to the output.

Logic Simulation: To see if a given circuit works correctly by exercising it on a computer. This replaces the time-consuming step of building a "breadboard."

Logic Symbol: *See* Logic Element.

Logical Primitive: Basic building blocks used in the electronic design process.

Low Controllability: Those faults that were not controlled to logic zero and logic one very often during the test sequence.

Low Observability: Those faults that were controlled to logic zero and logic one sufficiently, but had a low probability of being observed at a primary output.

LSI: *See* Large Scale Integration.

Manufacturing Test: The action of checking out a physical circuit by applying input to it and analyzing the output.

Mask: The photographic negative that serves as the master for creating circuit patterns in photoresist on wafers or for making thick-film screens or thin-film patterns.

Maze Router: A program that plots the paths of the electronic circuits in either semi-custom ICs or PCBs.

Meander: Any wiring that is over and above the minimum wiring needed to route a circuit.

Medium-Scale Integration: Standard digital integrated circuits having approximately 100 transistors.

Microprocessor: A digital computer that is implemented on a chip or set of chips. Some of the most popular microprocessor families are the Intel 8086, Motorola 68000, and National Semiconductor 32000.

Mirroring: To physically orient adjacent cells in an IC as mirror images to facilitate wiring.

Model: A representation of a particular electronic part or circuit that tells a simulator how that part or circuit should behave. Various model types include Hardware Models, Gate-Level Models, Functional Models, Behavioral Models, and VHDL Models.

Modeling: The process of describing the behavior of a hardware circuit or part—generally accomplished by combining primitives or by using a behavioral modeling language or a hardware description language. These models are used to perform logic simulation.

MOS (Metal Oxide Semiconductor): MOS integrated circuits are widely used throughout the industry since their original introduction in the mid-1960s. There are several major types of MOS technology in use today.

MSI: *See* Medium-Scale Integration.

Net: A signal path, node, or wire that connects two or more pins. (Also known as a signal network or signal net). All points on a net are considered at the same voltage or logic level.

Netlist: A netlist enumerates all of the parts in a circuit and how they are connected.

Netlister: A program that produces a file or set of files that describe a design, usually used to feed simulators or wire router programs.

Netlist Formatting: To extract the connectivity information of a circuit and put it in a form that can be read by another computer.

N-MOS (Negative–Metal Oxide Semiconductor): This technology became popular in the early 1970s because of its speed/performance characteristics. It is predominantly used by microprocessor manufacturers for devices such as the 68000, Z80, 8086, etc.

Node: (1) In a simulated circuit, a connection between pins. Sometimes referred to as a "net." (2) An Apollo workstation.

Nominal Delay: The average time a signal takes to propagate through an instance or net. The effect of an input change does not occur on the output until after the nominal delay duration.

Observability: The ability to detect a change in the state of a signal within a circuit by monitoring the primary outputs of the circuit. *See also* Controllability.

Oscillation Control: A way that simulators limit the number of times that a circuit can change state in a certain period of time.

PAL: *See* Programmable Array Logic.

Parallel Fault Simulation: A type of fault simulation that uses the same general approach as serial fault simulation, except that it simulates multiples of faults in parallel. Parallel fault simulation provides increased efficiency and speed over serial fault simulation, but is not as fast as concurrent fault simulation.

Partitioning: The logical grouping of electrical functions within a given set of hardware components.

Pin: A symbol's connection point between an external net and the internal circuitry of the component represented by the symbol.

PCB: *See* Printed Circuit Board.

PG Tape: The instrument used to transfer topological layout information to reticules or masks via pattern generators (usually used for creating thin-film hybrids, gate arrays, and semi- and full custom design).

PLA: *See* Programmable Logic Array.

Placement Area: The portion of a PCB where components can be placed.

Placement Class: Devices are grouped together according to the characteristics of their packaging. Devices in the same classes may be interchanged during the placement process.

Placement Protocol: A set of rules governing message exchange between two communication processes.

PLCC: Plastic leadless chip carrier. A square or rectangular package with leads on all four sides in which an integrated circuit chip can be mounted to form a surface mount component. Instead of leads it has pads around its perimeter for connection to the substrate.

PLD: A logic device that can be programmed after manufacture to perform a desired function. Programming is usually accomplished by blowing "fuses" in the device with carefully controlled patterns of electric current in a special piece of hardware called a logic programmer. These patterns of current are specified in a data file which in turn is generated by software designed for this purpose. The software generates the data according to a description supplied by the circuit designer.

P-MOS: While being the earliest MOS technology used for LSI, it is not the predominant MOS technology because it does not perform well at high speeds. The popularity of P-MOS is due to ease of manufacture and design. It is used primarily in low-cost consumer applications.

Point-to-point Wiring: An interconnecting technique wherein components are connected by wires routed between connecting points.

Possibly Detected Fault: A fault that occurs in a fault simulation when a known value in the fault-free simulation changes to an unknown value in the fault-inserted simulation.

Potentially Detected Fault: A fault that could not be identified as undetected with statistical analysis.

Preferred Direction: To maximize routing, a direction is normally assigned to each layer of the PCB. This is called the preferred direction.

PRF: Place, route, and fold. An early standard-cell automated layout tool.

Primary Input: An input pin of a circuit, to which you can apply input stimulus. For an IC device, the primary inputs are the input pins of the device. For a circuit board, the primary inputs are generally located on the edge connector.

Primary Output: An output pin of a circuit, from which you can read the circuit's response to input stimulus. For an IC device, the primary outputs are the output pins of the device. For a circuit board, the primary outputs are generally located on the edge connector.

Primitive: A simulation model that is specified at the lowest possible level; that is, a primitive does not contain other simulation models. Depending on the simulator, a primitive can be anything from a gate to a flip-flop to a RAM.

Printed Circuit Board (PCB): A common way of packaging electronic circuits in which the discrete components (resistors, capacitors, etc.) and/

or integrated circuits are mounted on a glass epoxy-type "board." Interconnections between the components are made by thin metal lines called etching. The PCB typically plugs into a chassis that contains other PCBs.

Programmable Array Logic (PAL): A rectangular array on a chip of programmable AND gates and fixed OR gates used to generate a group of control functions in sum-of-products form (a trademark of Monolithic Memories).

Programmable Logic Array (PLA): An array of gates that generates control functions in a sum-of products form. PLAs can be programmed to do different functions by external means.

Propagation Delay: The amount of time that it takes an input signal transition on a part or circuit to cause an output change.

Race: A simulation error that occurs when a bi-directional transfer device's input line and its enable line both change during the same simulation event.

Register Transfer Language (RTL): A commonly used language that allows the designer to describe the function of a design by defining the flow of information from one input or register to another register or output.

Rise/Fall Delay: The ability to assign a different value to the rise time (zero to one), and to the fall time (one to zero) of a device. This more accurately reflects the behavior of the actual device.

Routable Area: The area where it is acceptable to route interconnect wiring.

Router: A program that automatically determines the routing path for the component connections on a PC board or hybrid; also may be referenced in connection with the actions of a profiler.

Routing: The placement of interconnections on a printed circuit board.

RTL: *See* Register Transfer Language.

SA0 Fault: *See* Stuck-at Fault.

SA1 Fault: *See* Stuck-at Fault.

Sample Delay: The time between a clock signal transition and data capture. The delay should be long enough to let the circuit signals settle.

Sample Period: The time between data captures. The inverse of sample rate. The period is usually one-half the time of the clock period.

Sample Rate: The rate or frequency that a simulator captures data during a simulation. This rate is usually twice the clock frequency for the circuit.

Schematic Diagram: A diagram that represents an electrical circuit. Different-shaped symbols are used to represent the basic components of the circuit.

Schematic Library: A diagram of logic symbols established by the design engineer and forming an intelligent database that can be queried and tested for logic flaws.

Semiconductor: Materials used to build transistors. The fact that those materials could exist as either conductors or non-conductors led to the term semi-conductor.

Semi-custom ICs: Special integrated circuits that are designed in a shorter period of time. The benefits to the designer are that they receive the advantages of a custom IC without the inherent disadvantages of cost and design time. *See also* Gate Arrays and Standard Cells.

Sequential Logic: A machine whose next state is solely a function of its present state and its inputs.

Serial Fault Simulation: Serial fault simulation inserts into the structured design one fault mechanism at a time. Fault mechanisms are defined to be either a stuck-at-1 (SA1), a stuck-at-0 (SA0), or a shorted adjacent (SA) forced value at a device pin for an entire test program. The serial approach to fault simulation allows for only one fault mechanism to run through the entire test program. It is then compared against the fault-free simulation. If a difference is detected it is listed in the fault dictionary; otherwise it is considered undetected. Resimulation continues until all possible faults have been tried.

Serial Simulator: *See* Serial Fault Simulation.

Setup and Hold Primitive: An element inserted in a circuit that allows the checking of a prespecified condition.

Signal: The activity occurring on a net.

Signal Level: The logic state of a signal. The simulator can model three types of signal levels: logic 0, unknown (X), and logic 1. *See also* Unknown State.

Signal Strength: *See* Drive Strength.

Signal Value: *See* Signal Level.

Silicon Assembler: Advanced software that performs placement and routing functions for gate arrays and semi- and full custom circuits.

Silicon Compiler: A software design tool that facilitates IC design by use of a higher-level definition to translate the circuit into the layout specifications.

Simulation: A process whereby the entire database of a schematic is tested to ensure that the original circuit design is reflected in the schematic layout.

Simulation Kernel: *See* Back End.

Simulation Server: *See* Back End.

Single-Layer Metal Arrays: A type of gate array that employs a single layer of metal to make circuit interconnections.

Small-Scale Integration (SSI): Standard digital integrated circuits having approximately ten transistors.

SMC(SMD): A PCB term meaning surface-mount component (device). Any active or passive components designed to be physically mounted to the surface of a substrate instead of inserting leads into holes in the substrate.

Software Simulator: A program that verifies the functionality of a design prior to its being built.

SOIC: Small outline integrated circuit. A package in which an integrated circuit chip can be mounted to a surface-mount component. It is made of a plastic material that can withstand high temperatures, and has leads formed in a gull wing shape along its two sides for connection to the substrate footprints.

Solid-state: This term refers to the fact that a transistor can exist in a conductive or non-conductive "state" in one "solid" piece of material. Many different semi-conductor or solid-state technologies have developed over the last three decades with some becoming very predominant. The choice of semi-conductor technology is usually determined by the application, the ability to build, and the technical performance of that particular technology. Digital semi-conductor technologies basically fall into three large categories: MOS, TTL, and special.

SOS-MOS (Silicon-on-sapphire-MOS): This technology is becoming very popular among military vendors. Sapphire is used as the basic foundation (substrate) and the silicon is deposited onto that substrate. Diffusions are then produced using photo techniques as in the other technologies. SOS circuits are becoming popular for their high resistance to radiation, small size, and fewer imperfections.

SOT: Small outline transistor. A plastic leaded surface-mount component in which diodes and transistors are packaged.

Source Code: User-written (and therefore human-readable) programming statements before their translation into object code.

Spike: A simulation error that occurs when an input of an instance changes state within the propagation delay of the instance.

Spike Analysis: A simulator mechanism for detecting a potential spike and setting the signal value to unknown or as a potential error for a particular time interval.

SSI: *See* Small-Scale Integration.

Standard Cell: A process of creating an integrated circuit using predefined cells and connecting them in a custom arrangement. *See also* Gate Array.

States: A specific signal value that can be modeled during simulated circuit operation.

State Transition: A signal value change in a sequential machine, from one stable state to another.

State Value: One of twelve (depending on the simulator) specific signal conditions that the simulator can model. A state value consists of a signal level (1, 0, or X) followed by a drive strength (S, R, Z, or I). For example, 0Z is a logic 0 at high-impedance drive strength. *See also* Unknown State.

Static Fault: A fault due to logic levels as opposed to transients.

Stimulus: An input signal to a circuit that is intended to produce an output response.

Structured Design Methodologies: A way of designing using the top-down approach.

Stuck-at Fault: A type of fault in which a signal in the design is permanently held at a ground potential (stuck-at-zero, or SA0) or held at the positive power supply level (stuck-at-one, or SA1).

Substrate: The supporting material on which circuit elements are deposited.

Switch-level Model: A simulation model of a logic gate that allows a digital simulator to simulate the behavior of the transistors that comprise the gate. Each transistor is simulated as a switch, with no analog behavior taken into account.

Synchronous Logic: Circuits that use one or more clock signals to trigger the transition of logical signals.

TCE: Temperature or thermal coefficient of expansion, or coefficient of thermal expansion. The rate of expansion of a material when its temperature is increased. It is measured in ppm/C (parts per million per degree centigrade).

Testability: A measure of how easy it is to test a digital design; that is, a measure of the controllability and observability of the design. *See also* Controllability and Observability.

Testability Analysis: A method of measuring how well a design can be tested; that is, how well elements within a design can be controlled from its inputs and observed from its outputs.

Tester: A special-purpose hardware system, used to verify the correctness of a manufactured design. The tester administers predefined input test sequences and compares the results from the unit under test to known good-response values.

Test Pattern: Inputs and outputs that define a circuit's operation.

Test Sequences: *See* Test Pattern.

Test Vectors: *See* Test Pattern.

Thick film: A film deposited by a screen-printing process and fused by firing into its final form.

Thick-film Circuit: A microcircuit in which passive components of a ceramic-metal composition are formed on a suitable substrate by screening and firing.

Thick-film Hybrid Circuit: A thick-film circuit with active devices, usually chips, added on.

Thin Film: A film deposited on a substrate by an accretion process, such as vacuum evaporation, sputtering, or pyrolytic decomposition.

Thin-film Hybrid Circuit: A thin-film circuit with components, usually chips, added to a thin-film network.

Three-valued Logic Simulation: Simulation in which the values of the circuits are allowed to take on any of three values: zero, one, or unknown.

Time-critical Nets: A net where the exact delay time is crucial for proper operation.

Timestep: The smallest unit of time that the simulator can advance.

Timing Analysis: The process of checking the delays of signals within a circuit to see if the circuit operates properly.

Timing Assertion: Checking to see if signals are stable or changing.

Timing Verification/Verifiers: A software tool that checks for correct circuit operation at the specified speed.

Top-down Hierarchical Design: To design a circuit in such a way that the designer starts at a high abstraction level and breaks the circuit function into smaller blocks that each have more functionality included.

Trace: A line of etching on a printed circuit board.

Transfer Gate: A transistor switch that passes signals between two data points. Signals, in this case, do not refer to either power or ground.

Transistor: In digital circuit applications, a semiconductor device out of which logic gates are built.

Transport Delay: A type of propagation delay where the instance can accept input signals that transition faster than the instance's propagation delay. When the instance is using transport delay, the instance transfers these signal changes to its output without generating a spike warning. *See also* Inertial Delay, Propagation Delay, and Spike.

Trunk: A routing path.

Trunk Layer: The wiring layer on which wire segments run parallel.

TTL: Transistor-transistor logic. The type used by most digital circuits.

TTL Primitives: Building blocks that model some basic TTL components.

"Tunnel" Wiring Patterns: Predefined wiring segments that are prefabricated in polysilicon in some gate arrays.

Uncontrolled/Undetected Fault: A fault that has a detection probability of zero for that fault origin and is stuck-at-0 or is stuck-at-1. A pin is stuck-at-0 (stuck-at-1) if the pin-instance was never sampled to be logic-1 (logic-0).

Undetected Fault: A fault that has a detection probability of zero for that fault origin.

Uninitialized Node: A node in a simulator that has not yet reached its steady-state value.

Unit Delay: A simulation technique that sets all the delays of the elements to one time unit.

Unit Load: Any load configuration handled as a single item.

Unknown Level: A value for a signal that is neither one nor zero.

Unobserved/Undetected Fault: A fault that has a detection probability of zero for that fault origin and is unobserved stuck-at-1 (stuck-at-0). The pin-instance was sampled as a logic-0 (logic-1) but the fault effect was not propagated to a primary output where a tester could sense the difference between a good part and a faulty part.

User time unit: A user-defined period of time on which all time-related simulator commands are based (except where noted otherwise).

VDD: A term used in MOS circuits that describes the most negative power-supply voltage level.

Verification Test Sequences: The input stimuli used to validate a circuit's performance.

Verification Tools: *See* Design Validation Tests.

Very-Large-Scale Integration: An integrated circuit with more than 100,000 transistors.

VHDL: A hardware description language for modeling hardware and VHSICs (Very-High-Speed Integrated Circuits) in both a machine-readable and human-readable form. VHDL in this text refers to the IEEE Std 1076–1987 version.

Via: A hole for connecting two planes of a printed circuit board.

Via Spacing Rule: Defining the minimum separation between vias.

VLSI: *See* Very-Large-Scale Integration.

VSS: The most positive power-supply voltage level in a MOS circuit.

Waveform Input: To describe input by means of a waveform.

Wire Feature Spacing Rules: Describes the minimum separation between interconnections and other characteristics of the physical layout.

Wire List: A list of wires making only two connections each.

Wire Net: A subset of electrical connections in a logical net having the same characteristics and common identifiers.

Wire-ORing: To create a logical OR function by simply wiring gates together.

Wire Routing: The physical placement of the nets to achieve the desired connectivity.

Wire Rule: Restrictions that govern the placement of nets.

Wire Segment: Any straight portion of a net.

Wiring Channels: Portions of a board or chip used for routing the wires.

Wiring Congestion: The situation that occurs as the density of wiring increases.

Wiring Demand: A prediction as to the complexity of the wiring.

Workstation: The equipment used for CAD/CAE operations, consisting of a CRT and a computer.

Worst Case Delay: The acceptable delay values (longest and shortest).

Zero Delay: An attribute of a simulation technique in which all circuit-element delays are set at a value of zero time units (used to verify functionality).

Zero Delay Simulator: A simulator that has no delays for signal generation.

Zero/Unit Delay Simulator: A combination of the two types of simulators. The elements have zero delays while the feedback lines and memory elements have unit delays.

Index

AC test methodology
  characteristics of AC, 21
  propagation time, 153
  speed testing requirements, 153
Algorithm-based systems. *See* Expert systems
Analog cells, 169
Analog synthesis, 176
Application-specific integrated circuits (ASICs)
  advantages of use, 1–2, 8
  applications, 170
  future challenge, 176–177
  reasons for use, 9
ASIC types, 3
  advantages and disadvantages, 10–11
  compiled cells, 6
  full custom, 6–7
  gate arrays, 4–5
  programmable logic cell gate arrays, 7
  standard cells, 5–6
  trade-offs for costs, *11* (*table*)
Applications engineering, technical assistance, 24
Architectural level
  design creation example, *34* (*figure*)
  multilevel design, *5* (*figure*)
  top-down design, 35–36
Asynchronous events, 155
  pulse generators, *155* (*figure*)
Automatic test program generation (ATPG), 138

Behavioral language modeling (BLM), 81
Behavioral-level design, 37–39
  IEEE 1076–1987, 38
  VHDL, hardware description language *37–38* (*figure*)
Bidirectional pins, 50
Board-level tests, 163–166
  functional, 166
  in-circuit, 165
  timing, 124–125
Bottom-up design development, 57–60
  complexities of design, *58* (*figure*)
  cycle for design, 59
  hierarchical scheme, 59
  traditional method, 58

Boundary scan (JTAG), 159–161
  architecture, *160* (*figure*)
  IEEE P1149.1, 160
  JTAG architecture, *161* (*figure*)
  standards, 160
Built-in self-test (BIST), 158–159
Bus control, 48

Clock and control signal placements, 50
Combinational logic, 51
Compiled code modeling, 81–82
Conceptual-to-functional design, 61
  abstract descriptions, 61
  predesigned cells, 60
Concurrent design, 174
Connectivity, 77
Contractual requirements in fault simulation, 140–141
Control commands in logic simulation, 90
Counter and shift registers in test, 154–155
  breaking counter chains, *155* (*figure*)
Counter chains, partitioning of, 52
Critical path analysis (CPA), *122* (*figure*)
  advantages, 126–127
  board-level timing violations, 123
  limitations, 127
  race condition, 123
  setup or hold violations, 121
  slack, 121
  speed performance, 123
CrossCheck technology, 162

Datapath silicon compilers, 68. *See also* Silicon compilers
DC test methodology, 149–152
  burst, 149
  characteristics of DC, 21–22
  compare strobe time, 150
  conflicts, 151–152
  device under test (DUT), 149
  floating inputs, 152
  initialization problems, 152
  operating system, 149
  parameters, 148
  test cycle time, 149
  timesets, 150, *151* (*figure*)
  unit under test (UUT), 149
Design
  concept and specifications, 18–19
  creation, 33–56
  development steps, 9
  implementation, 24
  methodologies, 57–74
  readiness guidelines, 55
  rule checks, 54–55
  specifications, *19* (*table*)
Design kits. *See* Vendors for ASICs
Design parameters
  die size optimization, 44
  maximum speed, 44–46
  reliability, 46–51
  testability, 51–52
Design process, 17–31
  factors, 20–22
  flow, *18* (*figure*)
Design synthesis techniques, 63–65. *See also* Layout synthesis; Logic synthesis
  behavioral synthesis, 64
  complete automation, 63
Deterministic fault simulation, 131–132
Digital cells, 16

EDA. *See* Electronic design automation
Electrical optimization, 73
Electrical rule checks, 54–55
Electronic design automation (EDA), 41
  design environments, 173
  future of, 173–174
  in-house or design center, 30, 57
  research and development, 174, 176
  vendor choice, 171–173
  vendor requirements documents, 173
Embedded test matrix, 161–162
Expert systems, 66

False triggering, 48
Fanout and loading
  AOI configurations, 45
  buffers, 45
Fault dictionary, 133
Fault simulation, 77, 129–142
  acceleration techniques, 134–137
  advantages, 130
  commonly asked questions, 137–140
  coverage, 133–134
  design process, 130–131
  hardware acceleration, 137
  indeterminate fault, 138
  inputs, 132–133
  limitations, 141–142
  manually deleting faults, 138
  outputs, 133
  QuickFault fault simulator, *132 (figure)*
  testability, 130–131
  uses of, 129–130
Fault types, 131
Feedback loops in test, 154
  test reset, *154 (figure)*
Flattening or expanding the design, 79
Floating nodes, 47
Floorplanning, 44–45
Foundry/customer interface, 19–20; *See also* Vendors
Frameworks, 174
Functional design
  blocks, 40
  logic executions, 39
  simulation steps, *39 (figure)*
Functional optimization techniques. *See* Optimization techniques
Functionality evaluation of ASICs, 78–79
Functions/functionality, 22

Gajski chart, *36 (figure)*
Gate-level simulation accelerators, 175
Glitch generation
  gated clocks, 47
  race conditions, 47
Government requirements. *See* Contractual requirements
Graphical annotation, 133

Hardware acceleration in simulation, 99–102
  evaluation, 102
  hardware accelerator, *100 (figure)*
Hardware description language (VHDL), *37–38 (figure)*, 62–63, 82
  advantages, 63
  cell descriptions, 81, *84–85 (figure)*
  Department of Defense, 62
  IEEE-1076 Standards, 62
Hardware modeling systems, 175
Hardware simulation engines, 175
Hazards. *See also* Timing
  false reports, 119, 124
Higher-performance design tools, 174

IEEE P1149.1. *See* Boundary scan (JTAG)
Initialization and unknowns, 91
  inadequate initialization, *92 (figure)*
Input/output guidelines, 48–51
Iteration, 64

Joint Test Advisory Group (JTAG). *See* Boundary scan (JTAG)

Languages. *See also* Hardware description language (VHDL)
  behavioral, 60
  industry standard, 37
  register-transfer (RTL), 60, 63
Latch-up, 50
Layout parameter inputs, 74
Layout synthesis, 72–74. *See also* Optimization techniques, physical
Libraries
  basic cell, *43 (figure)*
  support, 171
  symbols for simulators, 43
  vendor, 43, 60
Local area network (LAN) acceleration, 134

Logfile, 133
Logic simulation, 77–107
  acceleration techniques, 96–102
  board-level, 107
  common mistakes, 104–105
  components, 79–90
  computer power, 94
  definition, 77
  limitations, 93
  purpose, 77
  simulator, *78* (*figure*)
  strategy, 102–104
  testability, 106–107
  tricks of the trade, 94–96
Logic synthesis, 66–67. *See also* Expert systems
  cycle description, *64* (*figure*)

Market statistics and predictions, 14–15
Mead, Carver, 67
Mentor Graphics Corporation, 29, 67
Min/max simulation, 77, *118* (*figure*)
  advantages of, 126
  common ambiguity removal, *119* (*figure*)
  disadvantages of, 121
  region of ambiguity, 118
  stuck-at faults, 121
  worst-case, 118
Mixed analog-digital ASIC
  cells, 169
  future of, 176
  synthesis, 65
Mixed-signal simulation tools, 176
Models, 80–86
  evaluation parameters, 86
  function, 80
  mixed-model simulation, 81
  timing, 81
  timing equations, *83* (*figure*)
  truth table, *82* (*figure*)
Module generators. *See* Silicon compilers
Module silicon compilers, 68. *See also* Silicon compilers
Murphy's law, 31, 139

N-channel transistors, 44, 47
NAND logic, 44
Netlist, 79, *80* (*figure*)
Non-return zero (NRZ), 150
Nonrecurring engineering (NRE) costs, 6, 21

Open architectures, 174
Optimization techniques, 64
  functional, 70–71
  performance, 71–72
  physical, 72–73

P-channel transistors, 44, 47
Packaging, 22
Physical/geometry design, manufacturing database, 41
Power analysis, 162–163
  bipolar and gallium arsenide, 163
  CMOS, 162
  uses of, 162
Power pads and drive current, 49
Primary inputs, 91
Primitives. *See* Simulation primitives
Production capability, 20

Quality assurance, 22

Random logic silicon compilers, 68. *See also* Silicon compilers
Realtime code execution, 175
Reliability design techniques, 46–51
Return one (R1), 150
Return zero (RZ), 150
Rule-based systems. *See* Expert systems
Rule/algorithm-based systems. *See* Expert systems

Sampling, 135
Scan in test design, 155–158
  advantages, 155
  automatic test pattern generator (ATPG), 158
  hypothetical scan system, *157* (*figure*)
  scan cells, *157* (*figure*)

synthesis, 158
techniques, 52
Schematic capture, 41–43, 79
components of, 42–43
NRE costs, 43
time-to-market window, 43
Schmitt trigger cells, 48
Sequential logic, 51
Sheets of primitives, 81
Silicon compilers, 37, 67–70
calibration steps, *69* (*figure*)
GENESIL floorplanning system, *72* (*figure*)
GENESIL screen shot, *70* (*figure*)
in design, 52–53
module generators, 67
types of, 68
Simulation, board and system, 174–175
Simulation primitives, 81
advantages and disadvantages, 81
boolean devices, 81
sequential devices, 81
Simulation states, 92–93
resolution table, *93* (*figure*)
Software prototype, 78
Spike generation, 46
Spikes and races, 114–115. *See also* Timing
Standards in test industry, 160
Statistical analysis in fault simulation, *135, 136* (*figures*)
Stimulus in logic simulation, 86–89
creating with a model, *96* (*figure*)
FORCE stimulus, *88* (*figure*)
graphical stimulus, *89* (*figure*)
high-level textual stimulus, *88* (*figure*)
list input, *87* (*figure*)
logfile input, *87* (*figure*)
Storage devices, 51–52
Structural design
detailed design, 40
logic synthesis techniques, 40
simulation steps, *41* (*figure*)
Stuck-at-1 and stuck-at-0, 131
Support, 22
choices, *25* (*table*)
Synchronous design techniques, 47
System on a chip, 33–34

Tabular (ASCII) tables, 133
Technology selections for ASICs, 11–14
BiCMOS two-input NAND, *12* (*figure*)
key characteristics, *14* (*table*)
Technology-independent design, 79
Test, 145–166
AC test methodology, 149–153
ASIC-tester interface, 147
checklist, 106–107
cost at each step, *164* (*figure*)
signal attenuation and distortion, *147* (*figure*)
technologies, emerging, 158–162
Testability, design for, 153–158
ad hoc concepts, 153–155
scan, 155–158
techniques, 51–52
Tile-based silicon compilers, 68. *See also* Silicon compilers
Timing, 111–125
board- and system-level, 124–125
circuit board performance, 111
conservative spike model, *117* (*figure*)
critical path analysis, 121–124
delay example, *112* (*figure*)
environmental effects, 114
hazards, 115–118
inertial and transport delay, *116* (*figure*)
layout information, 114
logic simulation, 114–115
min/max simulation, 118–121
parameters, 112–114
pin-to-pin delay, *113* (*figure*)
specialized analyzers, 118–124
traditional methods, 111–112
Timing analysis, 63
clocking violations, 63
critical paths, 63
setup violations, 63
Toggle test, 134

Top-down design, 60–61
  advantages, 60–61
  high-level requirements, 60
Turnaround time, 20–21

Unknowns. *See* Initialization and unknowns

Vectors, 77
Vendor and customer interfacing, 22
  design path, *22* (*figure*)
  responsibilities, *22* (*figure*)
Vendor for ASICs
  choosing, 24–31
  design kits, 27–29
  independent design, 79
  libraries, 26–27
  postdesign activities, 31
  support, 30
  track record, 26
Verilog hardware description language. *See* Hardware description language
Very-high-speed integrated circuit, 37
VHDL. *See* Hardware description language
VHSIC. *See* Very-high-speed integrated circuit

Workstations, 60
  ASIC vendor's use of, 171
  performance technology, 175
  simulation platform, 94
Worse-case timing, 77

Y chart. *See* Gajski chart

Zero delay simulation, 99